Supercritical Fluid Extraction and its Use in Chromatographic Sample Preparation

Supercritical Fluid Extraction and its Use in Chromatographic Sample Preparation

Edited by

S.A. WESTWOOD
Head of Analytical Methods
Horseracing Forensic Laboratory Ltd.
Newmarket

SPRINGER-SCIENCE+BUSINESS MEDIA, B.V

First edition 1993

Originally published by Chapman & Hall in 1993
Softcover reprint of the hardcover 1st edition 1993

Typeset in 10/12 pt Times New Roman by Pure Tech Corporation,
Pondicherry, India

ISBN 978-94-010-4958-0 ISBN 978-94-011-2164-4 (eBook)
DOI 10.1007/978-94-011-2164-4

A catalogue record for this book is available from the British Library

Library of Congress Cataloging-in-Publication data available

Foreword

by Professor D.E. Games, Mass Spectrometry Research Unit, University College of Swansea

Sample preparation can be viewed as occupying a Cinderella role in analytical science. However, the quality of sample preparation plays a key role in high quality analysis and deserves higher stature. In the past decade, there has been considerable interest in the use of supercritical fluid extraction (SFE) as an alternative to conventional procedures for the preparation of samples for analysis. The driving force for this development is the need for automated, simpler, faster, non-destructive and selective methods for extraction, preferably using non-toxic extraction media which are easily disposed of. Utilization of supercritical fluids for extraction fulfils these requirements because of their unique physical chemical properties and usually low toxicity. Selectivity can be achieved by suitable selection of pressure (density), temperature and modifier conditions which enable solvating power to be varied. The high diffusivity of supercritical fluids provides rapid sample penetration and extraction. Use of fluids with low critical temperatures enables extraction to be conducted under mild thermal conditions ensuring that thermally labile compounds do not decompose. The technique can be used off-line, and the extracts analysed by appropriate techniques, or it can be used on-line, by coupling with a variety of chromatographic techniques. These can then, if necessary, be coupled further with spectroscopic techniques, such as Fourier transform infrared, ultraviolet or mass spectrometry, to provide specific identification or structural information.

SFE is rapidly maturing from a research technique to one of application in a wide range of laboratories, and the appearance of this volume is timely. Dr Westwood has gathered together a group of authors who are all practitioners of SFE and its combination with the various chromatographic techniques covered. Hence, as well as providing the reader with a strong background in fundamental aspects of the use of SFE, practical aspects are also well covered and the reader is provided with a 'state-of-the-art' volume.

This decade will see SFE maturing into a technique widely used in the routine analytical laboratory as a replacement for many currently used extraction methodologies because it will be more cost-effective and will enable, through the provision of better quality extracts, better analytical data to be obtained. Automation of SFE is not yet fully developed; however, as the technique achieves wider acceptance, rapid development can be anticipated.

One can envisage, through the use of SFE on-line with the various chromatographic techniques and their interfacing to spectroscopic techniques, systems such as SFE-GC-UV-IR-MS and SFE-SFC-UV-IR-MS being commonly used where a sample is automatically placed in the extractor and there is no further contact. Since Baron Cagniard de la Tour discovered the supercritical phenomenon in 1822, a long time may appear to have elapsed before the widespread use of supercritical fluids. We can currently view ourselves as being in the Renaissance era of supercritical fluids, both for extraction and for mobile phases in chromatography.

D.E.G.

Contents

Contributors

Dr I.G.M. Anderson	British American Tobacco, Regents Park Road, Millbrook, Southampton SO9 1PE, UK
Professor K.D. Bartle	School of Chemistry, University of Leeds, Leeds LS2 9JT, UK
Dr A.A. Clifford	School of Chemistry, University of Leeds, Leeds LS2 9JT, UK
Mr N.J. Cotton	School of Chemistry, University of Leeds, Leeds LS2 9JT, UK
Dr S.B. Hawthorne	Energy and Environmental Research Center, University of North Dakota, Grand Forks, North Dakota, USA
Dr A.L. Howard	Department of Chemistry, Virginia Polytechnic Institute and State University, Blacksburg, VA 24061, USA
Dr M.W. Raynor	Department of Chemistry and Applied Chemistry University of Natal, Durban 4001, South Africa
Dr L.T. Taylor	Department of Chemistry, Virginia Polytechnic Institute and State University, Blacksburg, VA 24061, USA
Dr S.A. Westwood	Horseracing Forensic Laboratory Ltd., PO Box 15, Newmarket, Suffolk CB8 7DT, UK

1 Introduction to supercritical fluid extraction in analytical science

A.A. CLIFFORD

1.1 Introduction

Supercritical fluid extraction (SFE) is becoming an important tool in analytical science, and has seen rapid development in the last few years. Manufacturers are now producing instrumentation designed for routine application of the technique. It has the advantages, compared with liquid extraction, that (a) it is usually less expensive in terms of solvent cost and laboratory time, (b) the solvent is easier to remove, (c) pressure (as well as temperature and the nature of the solvent) can be used to select, to some extent, the compounds to be extracted, and (d) carbon dioxide is available, to be used as a pure or modified solvent, with its convenient critical temperature, its cheapness and non-toxicity. This book describes the principles and methods available for those considering using the technique for their analytical problems. This first chapter explains the basic principles of SFE, and starts by discussing some general points.

For SFE to be of use to the analytical chemist, it must be quantitative. To achieve this, a procedure needs to be developed for each application on the basis of a good understanding of the extraction process, in order to avoid a number of possible difficulties. In fairness to SFE, problems also occur in liquid extraction, but we have not been so aware of them, because the much slower techniques have not allowed the collection of sufficient data to properly evaluate them.

It is important in particular to understand the characteristic kinetics of SFE. Extraction by a supercritical (or any) fluid is never complete in finite time. It is relatively rapid initially, but there then follows a long tail in the curve of percentage extracted versus time, as shown in Figure 1.1. In a typical situation 50% is extracted in 10 min, but it may be 100 min before approximately 99% is extracted. It is not correct, therefore, to assume that extraction is essentially complete if it has been carried out for two consecutive equal periods of time and the second period produces only a fraction of the compound extracted in the first period. It is necessary, for every application, to carry out an experimental long extraction and study the results by the methods given below.

From the viewpoint of methodology, SFE is often classified as off-line or on-line. In off-line SFE, the sample is subjected to a flow of fluid, usually at

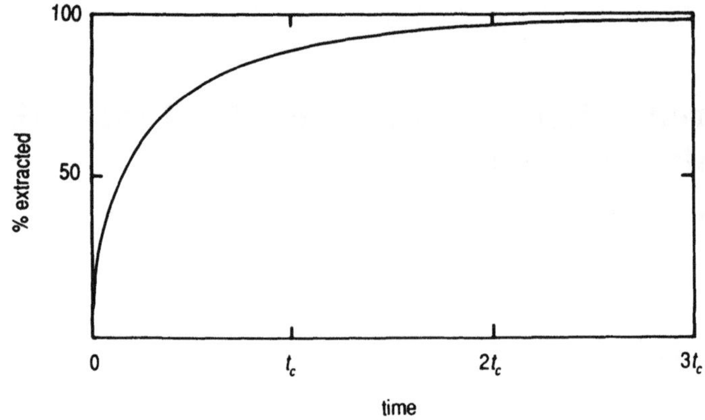

Figure 1.1 Theoretical curve of percentage extraction versus time.

constant temperature and pressure, and a series of samples collected at regular time intervals from the eluting fluid after depressurizing, by passing it through a solvent, for example. These samples are analysed later. In on-line SFE, the SFE instrument is coupled directly to the analytical instrument, as in SFE supercritical fluid chromatography (SFE-SFC), for example. Typically the sample is extracted by a flowing stream of fluid at a particular temperature and pressure for a certain length of time and the extract deposited, after depressurizing, on the front of an SFC column. The extraction is then stopped, while chromatographic analysis is carried out. The two processes are then repeated alternately. From the theoretical point of view, a better classification is into continuous and discontinuous SFE, and this classification is used in this chapter. In the examples given above, off-line was continuous and on-line discontinuous, but this need not necessarily be the case. For example, in on-line SFE-SFC, it is possible to arrange continuous extraction using two columns.

The discussions made below assume that ideal experimental procedures are used, meaning, for example, that the fluid pressure and temperature reach the desired values rapidly in comparison with the extraction times and that there is no significant 'hold-up' of solute in the extractor. Otherwise the results will not necessarily follow the models given closely, and extraction rates may depend on extraction cell volume, for example. In larger, non-analytical extractions, it is common for no material to be extracted for a significant period of time at the beginning of the process, during which the extraction vessel is being filled and brought up to temperature. However, it is possible in analytical SFE, using procedures such as are described in chapter 2, to get close to ideal conditions, so that the results may be interpreted by the methods given below.

1.1.1 Basic principles of SFE

Three interrelated factors influence recovery, as shown in the SFE triangle:

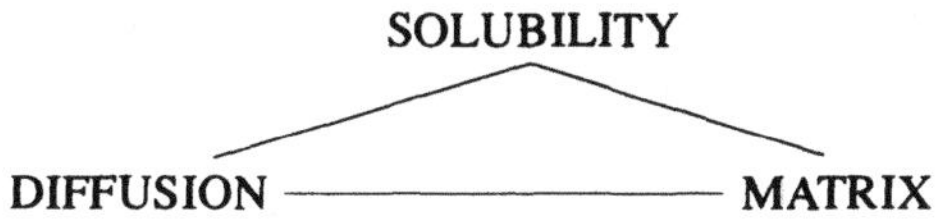

For extraction to be successful, the solute must, firstly, be sufficiently soluble in the supercritical fluid. This factor is especially important at the beginning of an extraction, when extraction is occurring at a higher rate. In many cases, this simply means being above 'threshold pressure', as illustrated, for example, in Figure 1.15. Solubility is therefore discussed in some detail towards the end of the chapter, although it is possible to obtain a crude indication of solubility from SFC retention. Control of solubility can allow stepwise extraction. For example, the bulk of the triglycerides may be extracted from peanut meal by CO_2 alone before aflatoxin analytes of interest are extracted by CO_2 containing methanol [1].

Secondly, the solute must be transported sufficiently rapidly by 'diffusion' from the interior of the matrix in which it is contained. The 'diffusion' process may be normal diffusion of the solute, or it may involve diffusion of the fluid into the matrix and perhaps subsequent replacement of solute by solvent molecules on surface sites. Often the precise process will not be known, but fortunately it can be successfully modelled as diffusion and given an effective diffusion coefficient. Henceforth, in this chapter diffusion is used to describe all possible processes for transport of the solute out of the matrix. The timescale for diffusion depends on the diffusion coefficient and the shape and dimensions of the matrix or matrix particles. Of these, the shortest dimension is of great importance, as the times depend on the square of its value. Values for this quantity of 1 mm or preferably less are usually necessary.

The third factor is that of the matrix (other than its effect on diffusion). Matrix effects mean that, although in many cases SFE will extract all of a particular compound in a sample, in some cases not all of a compound is 'extractable'; the rest being locked into the structure of the matrix, or too strongly bound to its surface. An example is the SFE of some additives and lower oligomers from polymers, where SFE can give much lower results than obtained by dissolving the polymer in a solvent, or using liquid extraction at higher temperatures, which swells the polymer. A similar situation occurs with an adsorbtive matrix, where, for example, the SFE of dioxins from fly-ash has been shown to be difficult, although it can be improved by adding acetone to the extraction cell [2].

Of these three factors, that of the matrix is the least well understood at present and a detailed scientific discussion cannot be given. In most cases, the analyst needs to quantify the total amount of a particular compound in a sample and needs to extract all the compound. A particular application has therefore to be investigated experimentally by: (a) varying the SFE technique used (e.g. the fluid) and testing for consistency; (b) comparing SFE with other methods; or (c) testing it with spiked samples (although spiked samples, even if aged, may not be truly representative of a real sample, especially for plant and animal tissue). The analyst may be able to develop a quantitative technique, but in some cases will not be successful. Liquid extraction techniques may also not be quantitative. For extraction from a polymer, chain entanglements make extraction by any method of molecules above a certain size inevitably less than complete. The amount 'extractable' under particular SFE conditions, even if less than 100%, may, however, be of interest, for example if one is concerned with migration of additives from polymers into foodstuffs.

Of the two remaining factors, solubility and diffusion, the latter is of more concern in analytical extractions. This is firstly because in most applications, the analyte is present in small quantities in the matrix and secondly a fluid and conditions will have been chosen in which the analyte is soluble at least to some extent. In any case, solubility is only important at the beginning of an extraction in a typical flow system, as a high rate of extraction occurs then and eventually the concentration of solute in the fluid is well below its solubility. This chapter, therefore first deals with a flow system in which it is assumed that the concentration of solute in the fluid is essentially zero (or alternatively the solubility of the solute in the fluid or the flow rate are essentially infinite). The effect of solubility limitation on the beginning of the extraction process is then discussed followed by a discussion of solubility in a supercritical fluid. Before treating SFE, however, we begin with an introduction to the nature of a supercritical fluid.

1.1.2 Pure and modified supercritical fluids

A pure supercritical fluid is a substance above its critical temperature and pressure. Above its critical temperature, it does not condense or evaporate to form a liquid or a gas, but is a fluid, with properties changing continuously from gas-like to liquid-like as the pressure increases. Figure 1.2 shows the phase diagram of a single substance. If we have liquid (L) and gas (G) in equilibrium and we move along the gas–liquid coexistence curve, TC toward C, increasing both the temperature and pressure, the liquid becomes less dense because of thermal expansion and the gas becomes more dense as the pressure rises. At the critical point, C, the densities of the two phases become identical and the distinction between the gas and the liquid disappears: the substance becomes a 'supercritical fluid'. The critical point is characterized by its two coordinates on the diagram, the critical temperature, T_c, and pressure, p_c. (There is also an associated critical density, ρ_c, and its inverse, the critical volume, V_c.) Thus above T_c the substance is described simply as a fluid, F. At lower pressures the fluid has gas-like properties: nitrogen gas in a cylinder is, in fact, a fluid. The region of interest for SFE is the hatched area above T_c and around or above p_c where densities, solubilities and other properties are intermediate between those of typical gases and liquids.

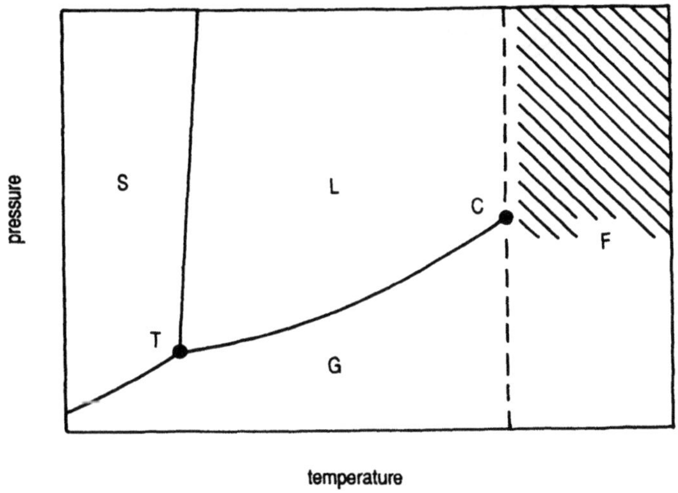

Figure 1.2 Schematic phase diagram of a single substance.

Table 1.1 shows the critical parameters of some compounds useful as supercritical fluids. One compound, CO_2, has so far been the most widely used, because of its convenient critical temperature, cheapness, non-explosive character and non-toxicity. Because the molecule is non-polar, it is classified as a non-polar solvent, although it has some limited affinity with polar solutes because of its large molecular quadrupole. Thus pure CO_2 can be used for many large organic solute molecules even if they have some polar character.

For the extraction and chromatography of more polar molecules, it is common to add modifiers or entrainers, such as the lower alcohols to CO_2, usually in small quantities. In such cases, it is important to be aware of the modifier–fluid phase diagram to ensure that the solvent is in one phase. For example for methanol–CO_2 at 50°C there is only one phase above 95 bar whatever the composition, but below this pressure, two phases can occur. The phase diagram for a binary mixture, such as methanol–CO_2, can be represented by a three-dimensional figure, whose axes are pressure, temperature and mole frac-

Table 1.1 Critical parameters of selected substances useful as supercritical fluids.

	T_c (°C)	p_c (atm)	ρ_c (10^3 kg m^{-3})
CO_2	31.3	72.9	0.47
N_2O	36.5	72.5	0.45
SF_6	45.5	37.1	0.74
NH_3	132.5	112.5	0.24
H_2O	374	227	0.34
n-C_4H_{10}	152	37.5	0.23
n-C_5H_{12}	197	33.3	0.23
Xe	16.6	58.4	1.10
CCl_2F_2	112	40.7	0.56
CHF_3	25.9	46.9	0.52

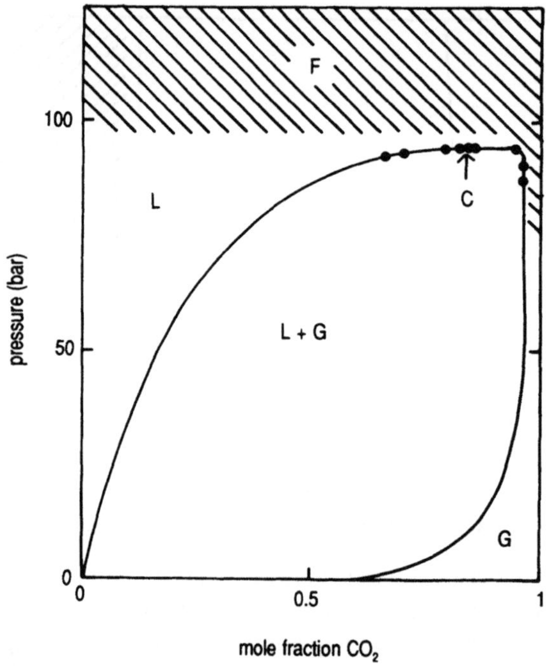

Figure 1.3 Semi-schematic phase diagram for methanol–CO_2 at 50°C. Points shown are from Brunner *et al.* [3].

tion. At a particular temperature, a cross-section through such a diagram is a two-dimensional pressure–mole fraction plot, of which an example is given for methanol–CO_2 at 50°C in Figure 1.3. This is a semi-schematic diagram, in which the phase boundary has been drawn through experimental points [3] and continued schematically to lower pressures. At very low pressures (which are not of importance in SFE) a single gaseous phase exists at all compositions, which are mixtures of CO_2 and methanol vapour. At intermediate pressures, both gaseous and liquid phases can occur, dependent on composition. At high mole fractions of CO_2 the mixture is gaseous (G), at high methanol concentrations it is liquid (L), and at intermediate compositions both phases exist (L + G). The L + G region reaches a (very flat) maximum in pressure at the critical point, C, for this particular temperature. Consider what happens to a mixture of the critical composition at a pressure below the critical pressure (where it will be in two phases) as the pressure is raised. The liquid will dissolve more CO_2, the gas will solvate more methanol, and the gas will increase in density. Eventually, at the critical point, the compositions and densities of the two phases will become identical. Thus above the critical pressure only one supercritical fluid phase, F, will exist. (One should mention that at very much higher pressures, of no concern in SFE, other phases, such as solids can occur.)

Thus to be under truly supercritical fluid conditions, the pressure needs to be above the critical pressure of the mixture for the particular temperature.

Table 1.2 Typical physical property values for gases, supercritical fluids and liquids.

	Density (10^3 kg m^{-3})	Viscosity (mPa s)	Self-diffusion coefficient (10^4 m^2 s^{-1})
Gas			
30°C, 1 atm	$(0.6–2) \times 10^3$	$(1–3) \times 10^2$	0.1–0.4
Supercritical fluid			
Near T_c p_c	0.2–0.5	$(1–3) \times 10^{-2}$	0.7×10^{-3}
Near T_c, $4p_c$	0.4–0.9	$(3–9) \times 10^{-2}$	0.2×10^{-3}
Liquid			
30°C, 1 atm	0.6–1.6	0.2–3	$(0.2–2) \times 10^{-5}$

However, in the context of SFE, where the proportion of modifier is often small, part of the gaseous phase is often considered as 'supercritical' as the pure gaseous component is above its critical pressure and temperature. Hence the hatched area in the figure is that usually loosely called 'supercritical'. It should be mentioned that, for both pure fluids and mixtures, many of the advantages of a supercritical fluid are possessed by liquids that are just sub-critical, and these are used in industrial processes, for example in the extraction of hops. The term near-critical is used to describe both situations and is preferred by some people. Again, although SFE is normally carried out by a one-phase fluid, because of possible experimental problems and inconsistent results, it is possible that a two-phase extraction may have an advantage in terms of the agitation of the matrix to be extracted.

SFE (and also SFC) take advantage of the fact that a supercritical fluid can have properties intermediate between those of a liquid and gas, and that these properties can be controlled by pressure. Table 1.2 shows some rather approximate typical values of important properties: density (this is related to solvating power), viscosity (related to flow rates) and diffusion coefficients (related to mass transfer within the fluid). One property advantage for SFE is that solubilities, and in particular the relative solubilities, of two compounds can be controlled via both pressure and temperature, making extraction selective to some extent. Other advantages are the relatively easy removal of the solvent and diffusion coefficients, higher than those of liquids, facilitating mass transfer in the extracting fluid. The disadvantage of using a supercritical fluid is that high pressure technology is involved. Although SFE and SFC are the two areas where supercritical fluids have been widely exploited, the use of these fluids is being extended into other areas, such as preparative SFC, chemical reactions, recrystallization and electrochemistry.

1.2 Extraction controlled by diffusion out of the matrix

1.2.1 Theoretical models for continuous extraction unlimited by solubility

We now consider the extraction of a matrix in a continuous flow of fluid, which is fast enough for the concentration of a particular solute to be well below its solubility limit. The rate-determining process is therefore the rate of diffusion out of the matrix. Most practical examples of extraction are complex, but it is found that simple models can account for the main features and lead to methods of treatment for the results of SFE. For these simple theoretical models, we assume an effective diffusion coefficient, D, and a particular geometry for the matrix and solve the appropriate differential equation (the Fourier equation) with assumed boundary conditions. These conditions are that the compound is initially uniformly distributed within the matrix and that as soon as extraction begins, the concentration of compound at the matrix surfaces is zero (corresponding to no solubility limitation). The solutions of the Fourier equation for various geometries are given by Carlslaw and Jaeger [4], in the context of heat conduction (where the same equation applies) and also by Crank [5], who has translated Carlslaw and Jaeger's equations into diffusion notation. Two simple geometries are discussed here: those of a sphere, which is applied to extraction of spherical particles as well as irregularly shaped powdered particles, and a slab with two infinite dimensions, which is applied to pieces of thin film.

The solution for a sphere, described as the hot-ball model because of the analogy of the mathematical solutions with those for a hot spherical object being dropped into cold water, is explained in a recent publication [6]. If the mass of solute in the matrix is m_0 initially and m after a given time, a plot of $\ln(m/m_0)$ versus time has the form given in Figure 1.4. It is characterized by a relatively rapid fall on to a linear portion, which corresponds to an extraction 'tail'. The physical explanation of the form of the curve is that the initial portion is extraction, principally out of the outer parts of the sphere, which establishes a smooth concentration profile across each particle, peaking at the centre and falling to zero at the surface. When this has happened, the extraction becomes an exponential decay.

The curve is characterized by two parameters: a characteristic time, t_c, and the intercept of the linear portion, $-I$, which has the value -0.5 (actually

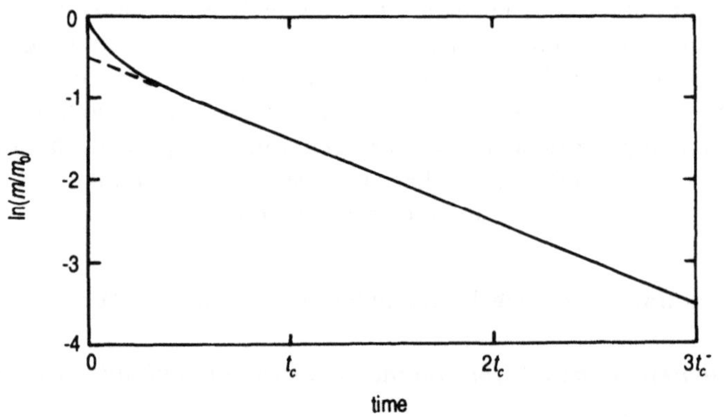

Figure 1.4 Theoretical curve for the hot-ball SFE model.

− 0.4977) for the sphere. The slope of the linear portion is $-1/t_c$ and the linear portion appears to begin at approximately $0.5t_c$. t_c is theoretically related to the effective diffusion coefficient out of the matrix, D, and the radius of the sphere, r, by the equation

$$t_c = r^2/\pi^2 D \tag{1.1}$$

The value of the effective D will usually not be known, although its order of magnitude may be commented on. Most measurements published for D are for true diffusion and for small molecules in relatively mobile solvents [7] and D is of the order of 10^{-5}. For systems of interest to SFE, D will be between 1 (for oils) and 4 (for solids) orders of magnitude below this value. Equation (1.1) shows a squared dependence on r and rationalizes the commonsense rule that for rapid extraction, matrix particles must be small. This may be achieved for solids by crushing or grinding and for liquids by coating on a finely divided substrate, spraying or mechanical agitation. For solid matrix particles with r of the order of 0.1 mm, typical values of t_c are between 10 and 100 min.

Figure 1.5 shows some experimental results for the extraction of 1,8-cineole from crushed rosemary at 50°C using CO_2 at 400 atm. Extraction was continued almost until exhaustion to allow the calculation of values of m and m_0. Similar curves are obtained for the extraction of five other major compounds from rosemary (α-pinene, camphor, camphene, borneol and bornyl acetate) and also for several other types of system [6]. The experimental results are consistent with the theoretical curve in that the points are close to a straight line after a time of approximately $0.5t_c$. t_c has a value of about 18 min in this case, which is obtained from the slope of the straight line portion (it is the time taken for the line to fall one natural log unit). However, the curve differs from the theoretical curve of Figure 1.4 in two respects. Firstly, the intercept is greater, and this is discussed in the next paragraph. Secondly, the curve does

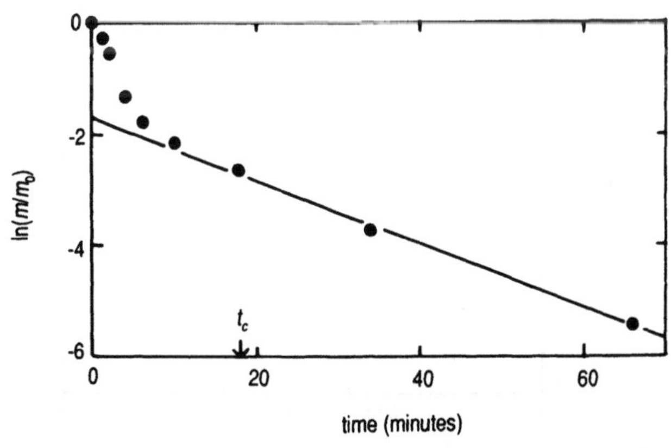

Figure 1.5 Off-line, continuous extraction of 1,8-cineole from crushed, dried rosemary with CO_2 at 50°C and 400 atm [6].

not fall as steeply from zero, and this is thought to be due to the effect of solubility limitation, which is discussed in a later section.

In general, the value of I is thought to depend on the particle shape and size distribution (for the former in particular the surface to volume ratio) and also the distribution of solute within the matrix particles (i.e. whether the solute is primarily located near the surface or in the interior of the particle). For a model system of spheres of the same size, with uniform solute concentration, it is 0.5. For real systems, values of approximately 2 are common and prediction of the values is not really possible. Thus, usually values of t_c and I can only be obtained by experiment. A continuous dynamic extraction, followed by treatment of the data by the methods described in this chapter, is therefore an important preliminary study in designing a routine quantitive analytical procedure. The experimental techniques required to carry out these studies are described in chapter 2.

The information in Figures 1.4 and 1.5 can also be given in terms of percentage extraction versus time, and this has been shown in Figure 1.1. For this figure, a more typical value of $I = 1.0$ is used, rather than the theoretical value of 0.5 for a sphere. As can be seen, the majority is extracted in a time of $0.5t_c$; 72% for the example given. Thereafter there is a long 'tail', corresponding to slow extraction; only a further 10% is extracted in the next period of $0.5t_c$.

Although the spherical model is adaptable to the irregular geometry of matrix particles, for extraction from a thin film of well defined geometry, a separate, though similar, study of a suitable model is desirable. In this case, our model would be that of an infinite slab of thickness L, on the basis that the surface dimensions of the film are far larger than its thickness. It is then necessary, as before, to solve the diffusion equation for the system with appropriate boundary conditions and the appropriate solutions are again given by Carlslaw and Jaeger [4]. Adaption of the published solutions leads to the curve of $\ln(m/m_0)$ versus time shown in Figure 1.6. The curve is similar to that

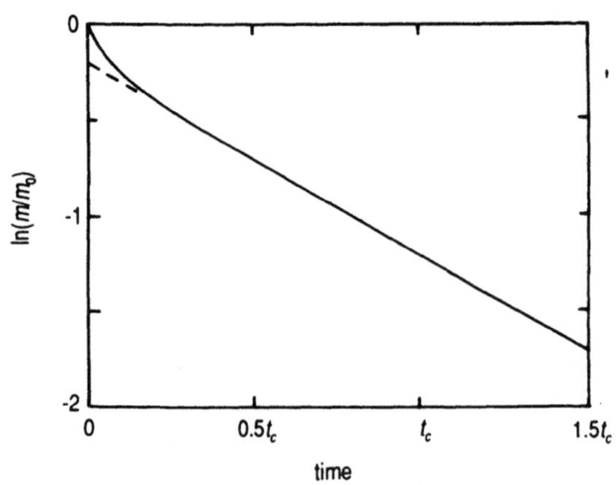

Figure 1.6 Theoretical curve for the simple thin-film SFE model.

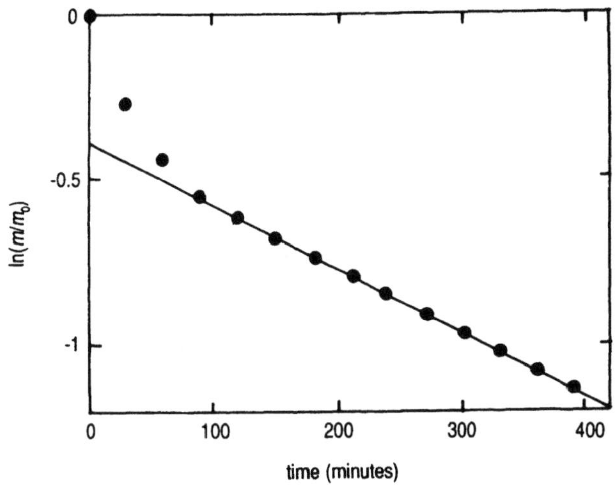

Figure 1.7 Off-line, continuous extraction of cyclic trimer from PET with CO_2 at 70°C and 400 atm [8].

for a sphere, with the curve falling more steeply initially, and later becoming approximately linear, with a slope of $-1/t_c$ where, in the case of an infinite slab,

$$t_c = L^2/\pi^2 D \tag{1.2}$$

However, it appears to fall more rapidly on to the straight portion than does the equivalent curve for a sphere, that is, after a time of approximately $0.25t_c$. Extrapolation of the linear portion of the curve to the time-zero axis, gives an intercept of -0.2100 that is, $I = 0.2100$, compared with a value for the sphere of 0.4977.

Qualitatively, the theoretical curve of percentage extracted versus time for an infinite slab is similar to that for a sphere, as shown in Figure 1.1, and exhibits the same long tail. 37% of the material is extracted during an initial period of $0.25t_c$. The time required to extract 99% of the material, however, is $4.4t_c$, that is, 17 times the time needed to extract the first 37%.

Figure 1.7 shows some experimental results for extraction from polymer film [8]. The sample was a film of poly(ethylene terephthalate) (PET), 1.2 mm in thickness, and extraction was carried out at 70°C with CO_2 at 400 atm; results are shown for the extraction of the cyclic trimer of ethylene terephthalate. Extraction was not carried out until exhaustion in this case as the time needed for this was impracticably long. Instead, the total amount of trimer extractable, needed for the calculation of values of m and m_0, was obtained using the extrapolation procedure described later. Figure 1.7 is a curve of the form of Figure 1.6 with a steeper portion falling on to a straight line after approximately 125 min. The slope of the straight line portion gives the result that $t_c = 506$ min. Thus the straight line appears to set in at $t_c = 0.25$, in agreement with the model. However, the value of I at 0.39 is above the theoretical value of 0.21 (as was obtained in the studies using the spherical model). A plausible explanation, in this case, is that a higher proportion of the oligomer near the surface is extractable under the conditions used. (It should be mentioned that the amount of cyclic trimer extractable under these conditions is considerably below that obtained by more rigorous extraction methods: an example of the existence of 'non-extractable' material in SFE.) From the slope obtained from Figure 1.7 and the thickness of the film, a value for the diffusion coefficient of the cyclic trimer in PET at 70°C can be obtained from the results to be $2.1 \times 10^{-13}\,m^2\,s^{-1}$. No literature value is available, but the result has the correct order of magnitude, by comparison with other diffusion coefficients in polymers [9].

In the case of the spherical model, the occurrence of an intercept below that of the theoretical value indicates either non-uniform distribution of extractable compound or irregular particle shape. In the case of extraction from a film of known geometry, the former is the only possibility, and so in this case it is worthwhile investigating the effect of non-uniform distribution on the theoretical results. A model distribution is required for such an investigation, and one of the simplest available for this purpose is an exponential fall-off in concentration from each surface. This is of the form $c_0 \exp(-x/a)$, where c_0 is a concentration, x the perpendicular distance in from the surface, and a a distance parameter giving the characteristic distance of the exponential fall-off. Figure 1.8 shows the concentration profile for a value for a/L of 0.3. The detailed solution is reserved for a later section and for the present it is sufficient to assert that the solutions are of the general form of Figure 1.6, but with the intercept becoming lower as a becomes smaller, that is, as the concentration falls off more rapidly from the surface. Table 1.3 give the values of I

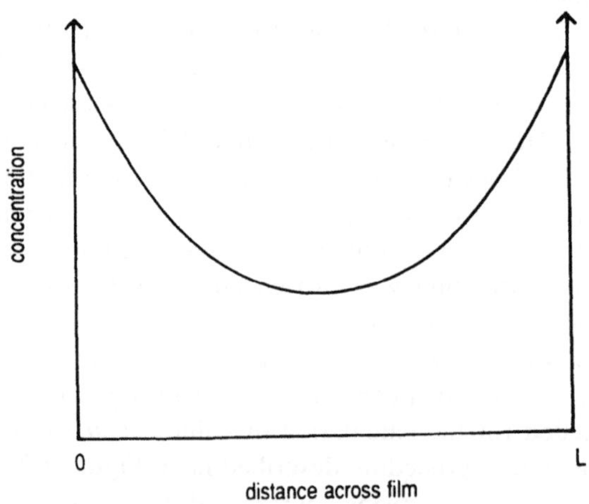

Figure 1.8 Model concentration profile across a film.

Table 1.3 Values for the intercept, I, for extractions from a film with non-uniform initial solute distribution for various values of the ratio of the distance parameter for the distribution, a, and the thickness of the film, L

a/L	I
∞	0.2100
1	0.2277
0.5	0.2779
0.3	0.3820
0.1	1.0103
0.05	1.6338
0.01	3.2199
0.005	3.9120
0.001	5.5215

expected for various values of a/L. A value of a/L tending to infinity corresponds to uniform concentration and gives the previous result. The value for I of 0.39 obtained with the results of Figure 1.7 is seen to correspond to a value for a/L of about 0.3, and this is the value for the concentration profile illustrated in Figure 1.8. Of course, the precise profile in the experimental film does not have to be of precisely this form, but Figure 1.8 indicates the extent of the predominance of extractable compound near the surface. The total cyclic trimer is probably uniformly distributed during manufacture, and the intercept value indicates that probably only the compound near the surface is more extractable.

1.2.2 Extrapolation of continuous extraction results

The exponential behaviour of the extraction after the initial period opens up the possibility of using extrapolation to obtain quantitative analytical information in a shorter time than would be required for exhaustive extraction. So far, only relatively simple models have been discussed, but it turns out that in all the more complex models discussed later for continuous extraction, the extraction tail is also exponential.

If extraction is carried out at least as long as the initial non-exponential period to obtain an extracted mass m_1, followed by extraction over two subsequent equal time periods to obtain masses m_2 and m_3, then it can be readily shown that m_0, the total mass in the sample, is given by

$$m_0 = m_1 + m_2^2/(m_2 - m_3) \tag{1.3}$$

It can be seen that if the value of m_3 is found to be very small and can be considered zero, that is, the extraction is almost complete after the second time period, the equation simplifies to

$$m_0 = m_1 + m_2 \tag{1.4}$$

as would be expected. If not, equation (1.3) may be used to obtain m_0, provided the difference between m_2 and m_3 is large enough compared with the errors in the two quantities. This is not too serious a problem, as usually the last term in equation (1.3) is small compared with m_1. This method is in the process of development, but it does appear that it could save in extraction time. Two polymer examples are given below, one involving pellets (nominally spheres) and the other a film.

In the case of extraction from polymers, there is an advantage in working with the original (rather than ground) sample pellets, as there is a danger, suggested by some of our experiments, that the results are affected by the grinding process (perhaps by loss of solute or a change in its extractability). However, a fairly exhaustive extraction of polymer pellets of a few millimetres in diameter is likely to take 80 h. The extrapolation procedure was therefore investigated for this type of system. Table 1.4 gives data for the extraction of 2,6-ditertiarybutyl-4-methylphenol (BHT) from standard polypropylene cylinders of 3 mm in both length and diameter, with additive concentrations known to within 1% w/w [6]. Although extraction was carried out for 8 h with only 57% of the additive extracted, an estimate of the final amount was made using equation (1.3) which is 5.2% below the given value. From the form of the curve of $\ln(m/m_0)$ versus time for this system, and calculations from the known diffusion coefficient for BHT in high density polyethylene [9], it can be deduced that the linear portion of the curve is not well achieved in the first extraction period. Thus the model is assuming the tail is falling off more rapidly than is in fact the case; hence the low result. If a better result were desired, this could be obtained with the sacrifice of a longer analysis time. It appears in this case that the great majority, if not all, the additive is extractable by SFE under the conditions used, perhaps helped by the fact that the BHT molecule is small.

Table 1.4 Extrapolation to obtain final quantities in the extraction of BHT (0.2% w/w) from 178.4 mg of standard polypropylene pellets using pure CO_2 at 50°C and at 400 atm [6]

Extraction times (min)	Weight extracted (μg)	Cumulative times	Weight extracted (μg)
0–20	7.1		
20–60	25.0		
60–120	45.7		
120–180	36.8		
180–240	26.8	0–240	141.4
240–300	16.4		
300–360	17.1	240–360	33.5
360–480	27.8	360–480	25.8
Total	202.7		
Given total	356.8		
Total from equation (1.3)			338.3
Difference between given total and equation (1.3): – 5.2%			

Table 1.5 Extraction to obtain final quantities in the extraction of cyclic trimer from PET film using pure CO_2 at 70°C and 400 atm [8]

Extraction times (min)	Weight extracted (μg)	Results predected from equation (1.3) (μg)
0–150	93.88	
150–270	19.08	
270–390	15.44	193.88
0–210	103.47	
210–300	13.61	
300–390	11.32	184.36
0–270	112.96	
270–330	8.14	
330–390	7.30	191.91

The second example is the extraction of cyclic trimer from the PET films illustrated in Figure 1.7 [8]. As the extraction was considered to be by no means exhaustive, the extrapolation procedure was used initially to estimate the total extractable oligomer. These calculations are shown in Table 1.5, in which three different sets of times were used. These gave results of reasonable agreement, with a total amount of 190 ± 5 μg ($\pm 3\%$) in the 2.739 g PET sample or 6.9×10^{-3} wt%. This is a considerably smaller percentage of oligomer than obtained by other methods. An explanation is that much of the oligomer is locked in and unextractable under the conditions of the extraction experiments, as discussed previously.

1.2.3 Treatment of discontinuous extraction data

It is a much more difficult problem to treat on-line analyses, where typically extraction is carried out for a certain period, following which extraction is stopped and analysis; for example, chromatography is carried out for a further period, these two processes then being repeated alternately several times. Qualitatively, this type of experiment is illustrated in Figure 1.9, which shows the development of the concentration profile of an extracted material across the film. The first period would be identical to off-line analysis, with initial uniform concentration, shown in (a) developing to the profile in (b). During the analysis period, when extraction ceases, diffusion of material continues to occur within the film down the concentration gradient, that is, towards the surface, producing the profile shown in (c). After extraction, the profile develops to (d), and so on. The extent of diffusion towards the surface during the analysis depends on the length of the analysis (non-extraction) period. This subsequently affects the amount extracted in a given time during the next extraction, as is shown by experiment.

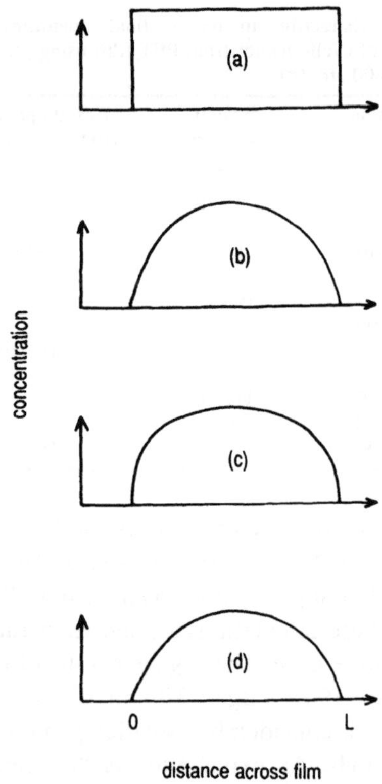

Figure 1.9 Schematic development of concentration profile during discontinuous extraction.

A mathematical analysis of this process [8], described in more detail in a later section, gives results in terms of a series of complex expressions, which must be numerically evaluated for a particular case. Treatment of results using these formulae are unsuitable for routine analysis. An experimental example is shown in Figure 1.10, which illustrates the results for the extraction of cyclic trimer from a similar sample of PET as was used for the results of Figure 1.7. Although the form of the curve of percentage extracted versus time is similar to that obtained for continuous extraction, consistent extrapolated results using equation (1.3) and a curve of the form of Figure 1.7 cannot be obtained from these results.

A simpler situation mathematically does occur if the periods for and between extractions are as long as the period necessary to reach the linear portion of the curve in a continuous extraction, that is, $0.25–0.5t_c$. Then a series of extractions each for the same period of time, interspersed with equal periods of non-extraction (not necessarily the same period as for extraction), would follow an exponential decay and the extrapolation formulae described in the last section are applicable in principle. Detailed equations are given in a later section. No experimental results have been treated in this way at present, however.

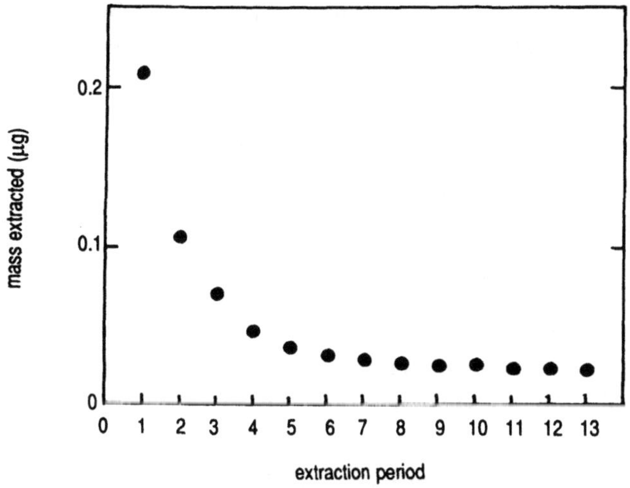

Figure 1.10 On-line discontinuous extraction of cyclic trimer from PET with CO_2 at 70°C and 400 atm [8], showing mass extracted during each extraction period.

1.2.4 Equations for continuous extraction models

In the last three sections, the mathematical content has been minimized to that essential for those wishing to make use of the models for treating data. In this section, more detailed descriptions of the equations and derivations for continuous extraction are given. In the following section, the equations for discontinuous extraction are derived and discussed.

The spherical or hot-ball model. We recall that the model consists of a solid sphere of radius r with a uniform initial concentration of a dissolved compound that is immersed into a fluid in which a zero concentration of the compound is maintained. Adaption of the published solutions for the differential equation (the Fourier equation), with appropriate boundary conditions [4], leads to the following equation for the ratio of the mass, m, of extractable compound that remains in the matrix sphere after extraction for time, t, to that of the initial mass of extractable compound, m_0.

$$m/m_0 = (6/\pi^2)\sum_{n=1}^{\infty} (1/n^2) \exp(-n^2\pi^2Dt/r^2) \tag{1.5}$$

where n is an integer and D the diffusion coefficient of the compound in the material of the sphere. Equation (1.5) may be simplified by defining, as was done in equation (1.1), a quantity t_c, which is a characteristic time for the extraction, to give, after expanding the summation,

$$m/m_0 = (6/\pi^2)\,[\exp(-t/t_c) + (1/4)\exp(-4t/t_c) + (1/9)\exp(-9t/t_c) + \cdots] \tag{1.6}$$

The solution is thus a sum of exponential decays, in which at longer times, the later (more rapidly decaying) terms decrease in importance and the first exponential term in the square brackets becomes dominant. This can be seen again if the natural logarithm of this equation is taken, after factorizing the term $\exp(-t/t_c)$ from the square bracket, to obtain

$$\ln(m/m_0) = \ln(6/\pi^2) - t/t_c + \ln[1 + (1/4)\exp(-3t/t_c) + (1/9)\exp(-8t/t_c) + \cdots] \quad (1.7)$$

The term $\ln(6/\pi^2)$ is equal to -0.4977, the final term in this equation equals $+0.4977$ at $t = 0$ and so, as required, at $t = 0$ $\ln(m/m_0)$ is also equal to zero. A plot of $\ln(m/m_0)$ versus time therefore tends to become linear at longer times, when the last term in equation (1.7) tends to zero, and $\ln(m/m_0)$ is given approximately by

$$\ln(m/m_0) = -0.4977 - t/t_c \quad (1.8)$$

Figure 1.4 is a plot of equation (1.7), and the straight line portion, which is continued to the time-zero axis as a dashed line, is a plot of equation (1.8).

The infinite slab model. Here the model is a rectangular slab of thickness L, whose other dimensions are infinite. As before, we assume initial uniform distribution of the compound to be extracted and a diffusion coefficient of D. In this case, as the slab is infinite, we must consider m to be the amount of the compound in a section of the slab, of given area, at time t and m_0 to be the amount in the same section at $t = 0$. Adaption of the appropriate solutions [4] gives

$$m/m_0 = (8/\pi^2)\sum_{n=0}^{\infty}(1/(2n+1)^2)\exp(-(2n+1)^2\pi^2 Dt/L^2) \quad (1.9)$$

where n is again an integer. As before, equation (1.9) may be simplified by defining, in equation (1.2), a characteristic time for the extraction, t_c and we obtain

$$m/m_0 = (8/\pi^2)[\exp(-t/t_c) + (1/9)\exp(-9t/t_c) + (1/25)\exp(-25t/t_c) + \cdots] \quad (1.10)$$

This is a similar sum of exponential decays, with the first exponential term in the square brackets becoming dominant at longer times, and this happens more rapidly than in the case of the spherical model. This equation gives, after factorization of $\exp(-t/t_c)$ from the square brackets, taking natural logarithms, and substituting the numerical value of $\ln(8/\pi^2)$,

$$\ln(m/m_0) = -0.2100 - t/t_c + \ln[1 + (1/9)\exp(-8t/t_c) + (1/25)\exp(-24t/t_c) + \cdots] \quad (1.11)$$

A plot of $\ln(m/m_0)$ versus time therefore again becomes linear at longer times and the nature of the equations is such that this linear portion is reached more rapidly than in the case of a sphere. The approximate equation for this model at longer times is

$$\ln(m/m_0) = -0.2100 - t/t_c \quad (1.12)$$

Figure 1.6 is a plot of equation (1.11), and the straight line portion, which is continued to the time-zero axis as a dashed line, is a plot of equation (1.12).

Extrapolation using the models. For both models, the equations approximated at longer times to equations (1.8) and (1.12), have the form, after taking exponentials, of

$$m = K \exp(-t/t_c) \quad (1.13)$$

where K is a constant for a particular extraction equal to $m_0 \exp(-0.4977)$ for a sphere and $m_0 \exp(-0.2100)$ for an infinite slab. Equation (1.13) is valid for a wide variety of models, including all the continuous extraction models described in this chapter. In fact, all solutions of the Fourier equation for continuous extraction, provided the boundary conditions are first order or less in concentration, are sums of exponentials approximating to a single exponential at longer times, however complex the geometry and however the initial concentration or diffusion coefficient depends on position within the matrix.

For the extrapolation procedure as described above, we require an initial extraction of an amount m_1, in a period of $t = 0 - t_1$, by which time the extraction should have approximately reached the linear portion of the curve of $\ln(m/m_0)$ versus time. There should then be two subsequent extractions, in equal periods, of amounts m_2 and m_3, these periods terminating at t_2 and t_3, respectively; that is

$$t_3 - t_2 = t_2 - t_1 \quad (1.14)$$

On substitution of the appropriate values at the end of each of the three extraction periods into equation (1.13), we obtain

$$m_0 - m_1 = K \exp(t_1/t_c) \quad (1.15)$$

$$m_0 - m_1 - m_2 = K \exp(t_2/t_c) \quad (1.16)$$

$$m_0 - m_1 - m_2 - m_3 = K \exp(t_3/t_c) \quad (1.17)$$

After dividing equation (1.16) by equation (1.15) and also equation (1.17) by equation (1.16), the following two equations are obtained:

$$(m_0 - m_1 - m_2)/(m_0 - m_1) = K \exp[(t_2 - t_1)/t_c] \quad (1.18)$$

$$(m_0 - m_1 - m_2 - m_3)/(m_0 - m_1 - m_2) = K \exp[(t_3 - t_2)/t_c) \quad (1.19)$$

From equation (1.14), the right-hand sides of equations (1.18) and (1.19) are identical and so the left-hand sides can be equated. After algebraic manipulation, we obtain equation (1.3), shown earlier; that is,

$$m_0 = m_1 + m_2^2/(m_2 - m_3) \tag{1.3}$$

Equation (1.3) is probably the simplest formula that can be used for extrapolation, but similar formulae, using different criteria for the collection of the data, can be derived using the methods given above.

Non-uniform distribution of extractable compound in a slab. Here the extractable compound in an infinite slab is considered to be distributed not uniformly, but with concentration falling off exponentially from the surface. The concentration is a function of the distance x along a coordinate perpendicular to the slab surface. If the slab surfaces are located at $x = 0$ and $x = L$, then the concentration is initially given by

$$c(x,0) = c_0[\exp(-x/a) + \exp(-(L-x)/a)] \tag{1.20}$$

where a is a distance parameter, as shown in Figure 1.8. The average concentration in the slab at time zero, obtained by integrating this expression with respect to x between 0 and L and dividing by L, is

$$c_{av}(0) = (2c_0/L)\,[1 - \exp(-L/a)] \tag{1.21}$$

Adaption of the appropriate heat conduction equation [4] gives the following integral equation for determining the concentration profile at a later time from the initial concentration profile, when the surfaces of the infinite slab are maintained at zero concentration by continuous extraction unlimited by solubility:

$$c(x,t) = (2/L)\sum_{n=1}^{\infty} \sin(n\pi x/L)\exp(-n^2t/t_c)\int_0^L c(x',0)\sin(n\pi x'/L)\,dx' \tag{1.22}$$

where t_c is defined by equation (1.2) and n is an integer. Substitution of $c(x,0)$ from equation (1.20) and integration leads to an expression for $c(x,0)$. This expression is then further integrated with respect to x and divided by L to give the average concentration across the slab at t, as follows:

$$c_{av}(t) = 8c_0(1 + \exp(-L/a))$$
$$\times\sum_{n=0}^{\infty}[1/((L/a)\,2 + \pi^2(2n+1)^2)]\exp(-(2n+1)^2t/t_c) \tag{1.23}$$

As it is true that

$$m/m_0 = c_{av}(t)/C_{av}(0) \tag{1.24}$$

m/m_0 can be obtained by dividing equation (1.23) by equation (1.21) to be given by

$$m/m_0 = (4a/L)\,[(1+\exp(-L/a))/(1-\exp(-L/a))]$$

$$\times \sum_{n=0}^{\infty} [1/((L/a)^2 + \pi^2(2n+1)^2)]\,\exp(-(2n+1)^2 t/t_c) \qquad (1.25)$$

This equation is of the same type as equation (1.9), derived for uniform initial concentration distribution, with a sum of exponential decays, of which the first term becomes dominant at longer times, that is $\ln(m/m_0)$ becomes linear with time at longer times. Also it can be readily shown that in the limit of very large values of a/L, which correspond to uniform distribution of concentration initially, equation (1.25) becomes identical with equation (1.9), as required. However, for finite values of a/L, and especially for values less than unity, the higher terms in the sum of exponentials fall off less rapidly in importance than was the case for uniform initial distribution. This results in the initial rapid extraction being of greater importance and a lowering of the intercept of the linear portion.

The value of the intercept, $-I$, may be found from the natural logarithm of the coefficient of the first term in the series (n=0); that is,

$$-\mathrm{I} = \ln\{(4a/L)\,[(1+\exp(-L/a))/(1-\exp(-L/a))]\,[1/((1+a\pi/L)^2)]\} \qquad (1.26)$$

Some values of I for various values of a/L are given in Table 1.3.

1.3 Equations for discontinuous extraction

In this section, equations for discontinuous extraction are developed for a particular case: the infinite slab with a uniform initial concentration distribution and with no solubility limitation. Equations for other geometries may be obtained by similar methods, but this particular example illustrates the main features of discontinuous extraction. We consider a series of N extractions, for periods of time $e_1, e_2, \ldots, e_N$, alternated with N–1 periods, $r_1, r_2, \ldots, r_{N-1}$, in which no extraction takes place. It will be necessary to calculate the development of the concentration profile along the direction perpendicular to the slab surfaces, $c(x, t)$, during these periods. During extraction, this may be done with the aid of equation (1.22). For the periods where there is no extraction, the following equation is adapted from Carlslaw and Jaeger [4], which describes the development of $c(x, t)$, under conditions where there is no flux of compound through the slab surfaces:

$$c(x,t) = (1/L)\int_0^L c(x',0)\,\mathrm{d}x'$$

$$+ (2/L)\sum_{n=1}^{\infty} \cos(n\pi x/L)\,\exp(-n^2 t/t_c)\int_0^L c(x',0)\cos(n\pi x'/L)\,\mathrm{d}x' \qquad (1.27)$$

where t_c is defined by equation (1.2) and n is an integer. Starting with uniform concentration, c_0, which is also equal to the average concentration, $c_{av,0}$, equations (1.22) and (1.27) are repeatedly applied for the appropriate periods of time. Also after each extraction period, the concentration is averaged over the slab by integration to give a value of $c_{av,N}$. This quantity is then divided by $c_{av,0}$ to give, by virtue of equation (1.24), a value of m/m_0 after the Nth extraction, $(m/m_0)_N$. The integrations necessary for this procedure are straightforward, but regrettably each application of the equations results in an additional summation being introduced into the formulations. The results can be represented by the following general formula:

$$(m/m_0)_N = (8/\pi^2)^N \sum_{i_1=0}^{\infty} \cdots \sum_{i_N=0}^{\infty} \sum_{j_1=1}^{\infty} \cdots \sum_{j_{N-1}=1}^{\infty} \prod_{k=1}^{N} \exp\{-(2i_{k+1})\, e_k/t_c\}$$
$$\times \frac{1}{(2i_1+1)^2} \prod_{l=1}^{N-1} \left[\frac{1}{(2i_{l+1}+1)^2} + \frac{(2i_{l+1})^2}{(2i_{l+1}+1)^2 - (2j_l)^2} - \frac{2\exp\{-(2j_l)^2 r_l/t_c\}}{(2j_{l+1}+1)^2 - (2J_l)^2} \right] \quad (1.28)$$

This formula is obviously unsuitable for routine analytical application, but it may be simplified if it is arranged that the experimental periods, e_i and r_j, are long enough, compared with t_c, so that only the first term in each sum is important. Comparison with the situation for continuous extraction indicates that the periods should be at least $0.25t_c$ for this approximation to be valid. In this case, equation (1.28) becomes

$$(m/m_0)_N = (8/\pi^2)^N \left[\prod_{k=1}^{N} \exp(-e_k/t_c) \right] \left[\prod_{l=1}^{N-1} [(1 + (2/9)\exp(-r_l/t_c)] \right] \quad (1.29)$$

If all the extraction periods are a fixed value, t_e, and all the periods between extraction are also a fixed time, t_r, the equation simplifies further to

$$(m/m_0)_N = [(8/\pi^2)^N \exp(-t_e/t_c]^N\, [1 + (2/9)\exp(-t_r/t_c)]^{N-1} \quad (1.30)$$

Thus, after the first extraction, the value of m/m_0 falls exponentially with the number of extractions and this can be seen by taking the natural logarithm of equation (1.30) to give

$$\ln[(m/m_0)_N] = -\ln[1 + (2/9)\exp(-t_r/t_c)]$$
$$+ \{\ln(8/\pi^2) - t_e/t_c + \ln[1 + (2/9)\exp(-t_e/t_c)]\}\, N \quad (1.31)$$

A plot of $\ln\{(m/m_0)_N\}$ against the number of extractions, N, should therefore give a straight line with an intercept of $-\ln[1 + (2/9)\exp(-t_r/t_c)]$. Experience with continuous extraction indicates that the intercept may be different from this value for various reasons. However, equation (1.31) looks promising for the analysis of discontinuous extraction data and indicates that the extrapolation procedure of equation (1.3) may be applicable.

1.4 Extraction controlled by both solubility and diffusion

1.4.1 The effect of solubility limitation on continuous extraction

So far it has been assumed that the solubility of the solute in the supercritical fluid is essentially infinite and diffusion out of the matrix has controlled the rate of extraction. To illustrate the interaction between solubility and matrix diffusion considerations, extraction of a solute from a stationary matrix in a cell by steady flow of the fluid is now discussed. Flow rates are now slow enough so that, at the beginning of the extraction, the concentration of the solute in the fluid is an appreciable fraction of its solubility, which varies with pressure. Figure 1.11 shows theoretical plots of the type of Figure 1.4, but now with solubility limitation, at a number of pressures, but constant flow rate. Curve 1 is at the lowest pressure and curve 5 at the highest. Curve 6 is that for infinite solubility, i.e. the earlier model illustrated in Figure 1.4, for comparison. The effect of solubility limitation is to reduce the high rate at the beginning of the extraction and also to reduce the slope of the linear portion, with the extent of these effects increasing as the pressure falls and the solubility decreases.

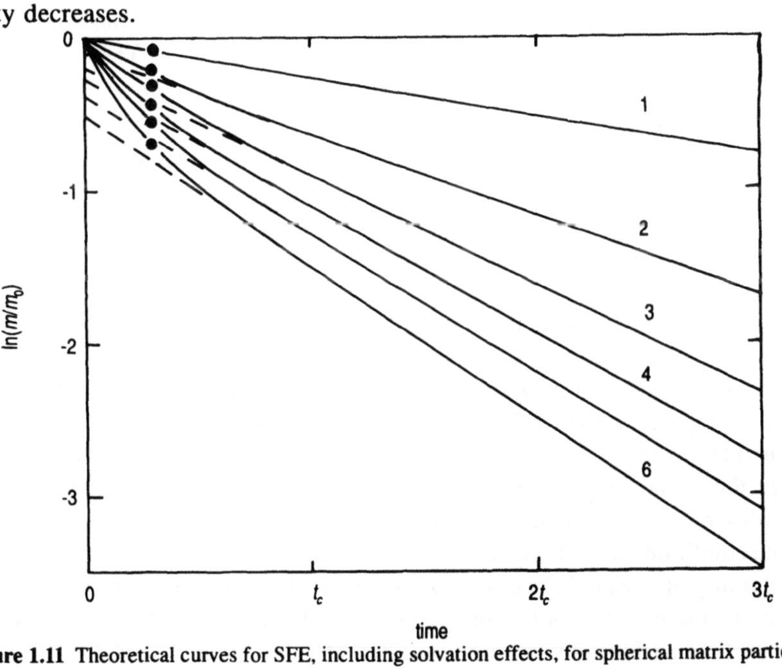

Figure 1.11 Theoretical curves for SFE, including solvation effects, for spherical matrix particles.

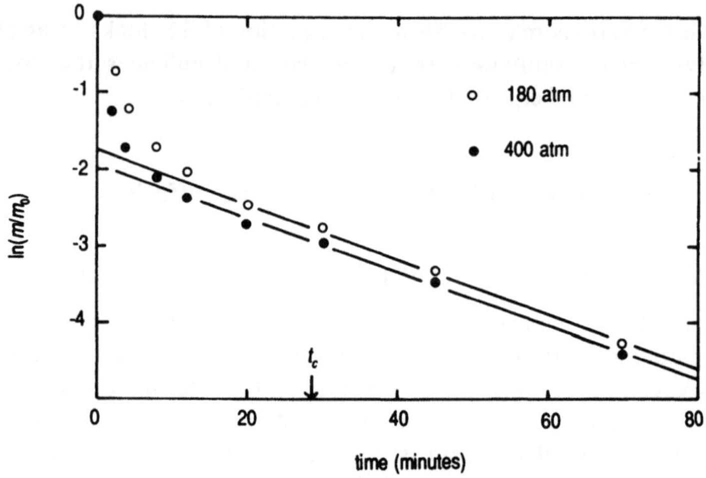

Figure 1.12 Off-line, continuous extraction of phenanthrene from railroad soil by CO_2 at 50°C and two different pressures at constant flow rate [6].

Figure 1.12 shows the extraction of phenanthrene by supercritical CO_2, at 50°C and with a constant flow rate from railroad-bed soil at two extraction pressures; 400 atm (filled circles) and 180 atm (open circles) [6]. In both cases, the curve of $\ln(m/m_0)$ versus time has the form of the hot-ball model, with an initial steep fall becoming linear at approximately $0.5t_c$. However, at 180 atm the curve falls less steeply initially and the linear portion is displaced upwards, although within experimental error the slope is not changed. Thus some of the features of the theoretical curves of Figure 1.11 are seen. They can be explained by the difference in solubility at the two pressures. At 50°C the saturated mole fraction of phenanthrene in CO_2 is 0.0015 at 180 atm compared with 0.003 at 400 atm [10]. Although the two curves in Figure 1.12 appear very similar, at short times the amount extracted is different at the two pressures. After 4 min 82% is extracted at 400 atm, whereas 71% is extracted at 180 atm. At even lower pressures, the amount extracted would be expected to be even lower and the initial fall would be further flattened out.

Figure 1.13 gives the extraction curve for carbazole from railroad soil under the same conditions and at 400 atm where its solubility in CO_2 is about an order of magnitude lower than phenanthrene (i.e. 0.00025 in mole fraction terms compared with 0.003 for phenanthrene at 50°C and 400 atm, the former value having been obtained by a small extrapolation of published results [11]). Here the initial fall has been almost completely flattened and only 25% has been extracted after 4 min, compared with 82% for phenanthrene under the same conditions. Although solubility limitation must be an important factor in explaining the form of the curve for carbazole, it is possible that a partial cause could be the fact that carbazole, being somewhat polar, has been leached from the particle surfaces by solution in rain water.

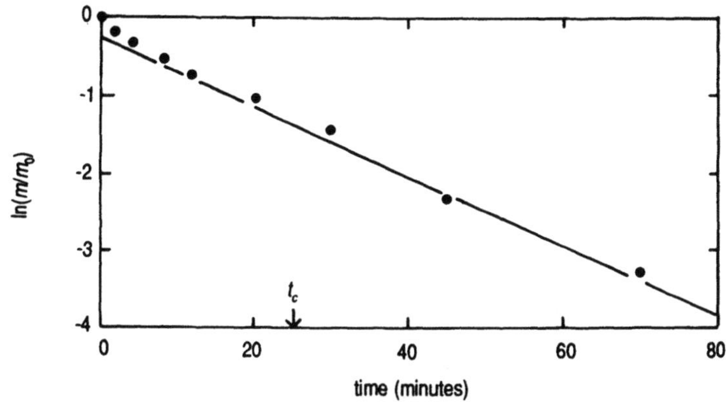

Figure 1.13 Off-line, continuous extraction of carbazole from railroad soil by CO_2 at 50°C, 400 atm and the same flow rate as in Figure 1.12 [6].

A more complete study of the effect of pressure and solubility on the kinetics of extraction has recently been carried out [12] and is illustrated in Figure 1.14. This figure shows the extraction of Irgafos 168 (tris (2,4-ditertiarybutylphenyl) phosphite) from ground polypropylene at 70°C. Experiments were carried out at a constant flow rate and pressures of 75, 105, 175, 200 and 400 bar. The highest point at each time is for 75 bar, the lowest for 400 bar, with results for intermediate pressures moving progressively downwards with

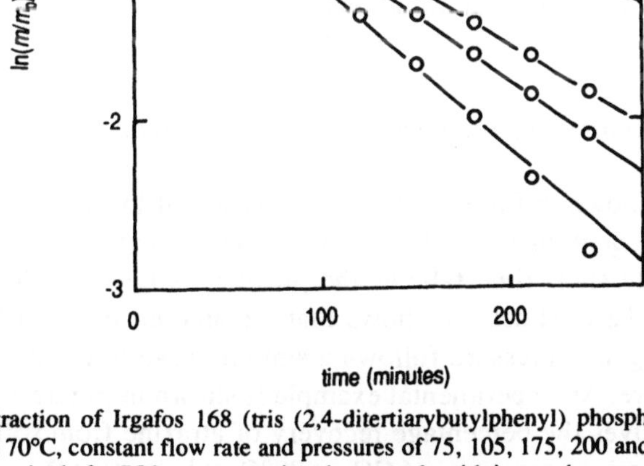

Figure 1.14 Extraction of Irgafos 168 (tris (2,4-ditertiarybutylphenyl) phosphite) from ground polypropylene at 70°C, constant flow rate and pressures of 75, 105, 175, 200 and 400 bar. At each time, the highest point is for 75 bar, progressing downwards with increasing pressure to the lowest point, obtained at 400 bar [12].

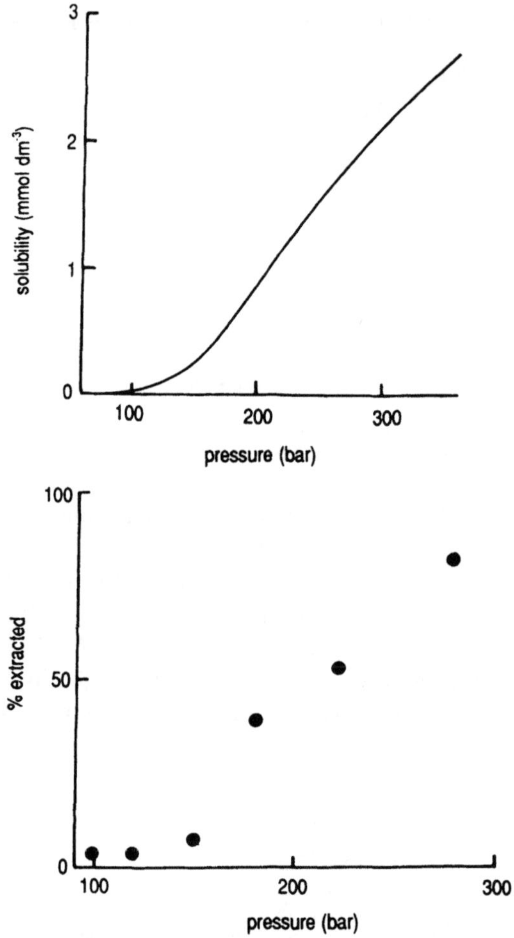

Figure 1.15 Percentage recovery of atrazine from soil by extraction with CO_2 at different pressures after 15 min at 80°C and constant flow rate, compared with predicted solubility at the same temperature [13].

increasing pressure. These curves mirror the predictions of Figure 1.11 qualitatively.

The points shown in Figure 1.11 are the values of $\ln(m/m_0)$ at $t = 0.3t_c$ with t_c defined by equation (1.1). These show that the amount recovered after a given relatively short time falls as the pressure is lowered. A more detailed discussion in the next section shows that the amount recovered after a short time plotted against pressure follows a similar curve to the plot of solubility against pressure. An experimental example is shown in Figure 1.15, where the lower graph gives the percentage recovery of atrazine from soil after 15 min extraction at various pressures of CO_2 at 80°C and constant flow rate [13]. The upper graph shows the predicted solubility of atrazine in CO_2 at the same temperature as a function of pressure and it can be seen that the two plots have similar shapes.

1.4.2 Models for solubility limitation

In previous models it has been assumed that the concentration of solute in the fluid adjacent to the matrix surfaces is zero and consequently the concentration of solute at the surfaces of the matrix particles or films is also zero. The equations were obtained by using $c(0, t) = 0$ as a boundary condition during extraction. We now assume that the concentration of solute in the fluid, c_f, is finite and is determined by the rate of loss of solute out of the matrix and the volume rate of flow of fluid, F. The appropriate equation is

$$-AD(\mathrm{d}c/\mathrm{d}x)_{x=0} = Fc_f \tag{1.32}$$

where A is the surface area of the matrix and x is the distance perpendicular to the surface, increasing from negative to zero at the surface. The term on the left-hand side of equation (1.32) is therefore, by Fick's first law of diffusion, the rate of loss of solute from the matrix. This is equal at the steady state, which is assumed to be rapidly established, to the rate of transport of solute away from the matrix, which is given by the right-hand side of the equation.

It is now assumed that the concentration in the fluid is proportional to the concentration in the matrix at the surface, and that the partition coefficient is proportional to the solubility of the solute in the fluid. The latter assumption in physical terms means that it is asserted that the solvation properties of the fluid for the solute affects the solubility and the partition coefficient equally. This latter assumption is not vital for the equations to be developed, if the model is described as one for 'solvation limitation' rather than 'solubility limitation'. However, if both assumptions are made, we obtain

$$c(0, t) = (1/KS)\, c_f \tag{1.33}$$

where K is a constant for a particular matrix, temperature and solute, with units of inverse concentration, and S is the solubility in units of concentration. Combining equations (1.32) and (1.33) gives

$$-(\mathrm{d}c/\mathrm{d}x)_{x=0} = hc(0, t) \tag{1.34}$$

where

$$h = KSF/AD \tag{1.35}$$

Equation (1.34) is the boundary condition that is used in conjunction with the diffusion equation to obtain the equations for solubility limitation. It can be seen that if h becomes very large, compared with the inverse of the distances involved, then $c(0, t)$ becomes close to zero and the simple models developed earlier apply. This occurs as expected, from equation (1.35), if the solubility and flow rate are high and the diffusion coefficient, that is, the rate of diffusion out of the matrix, is low. Solubility limitation therefore occurs as S or F and therefore h fall in value.

The infinite slab. Adaption of the appropriate heat conduction equations [4], which in this case correspond to Newtonian heat loss from a slab, gives the following equation for extraction from an infinite slab with uniform initial concentration distribution:

$$m/m_0 = 8 \sum_{n=1}^{\infty} (hL/a_n)^2/[hL(hL+2)+a_n^2] \exp(-a_n^2 Dt/L^2) \quad (1.36)$$

where m amd m_0 are the amounts in a section of the slab cut perpendicular to any given area of the surface and hL are the roots of the equation

$$a \tan(a/2) = hL \quad (1.37)$$

These roots are tabulated or are available from computer library routines and, in the limit of large values of hL are given by

$$a_n = (2n-1)\pi \quad (1.38)$$

For very large values of hL, therefore, it can be readily shown that equation (1.36) approaches equation (1.9), that is, the equation for an infinite slab in a fluid with zero solute concentration, as expected. Equation (1.36) is not discussed further, except to say that it gives a series of curves of the general form of Figure 1.11. The corresponding equation for spherical geometry is now developed and discussed in more detail, as it has greater applicability. The discussion for slab geometry is analogous.

The spherical model. Adaption, as before, of the appropriate heat conduction equations [4], which again correspond to Newtonian heat loss, gives, after integration over the sphere to obtain the average concentration and some algebraic manipulation, the following equation for extraction from a sphere with uniform initial concentration distribution:

$$m/m_0 = 6 \sum_{n=1}^{\infty} (hr/a_n)^2/[hr(hr-1)+a_n^2] \exp(-a_n^2 Dt/r^2) \quad (1.39)$$

where a_n are, in this case the roots of the equation

$$a \cot(a) = 1 - hr \quad (1.40)$$

Again these roots are tabulated or are available from computer library routines and, in the limit of large values of hr are given by

$$a_n = n\pi \quad (1.41)$$

For these very large values of hr, it can be shown by substitution that equation (1.39) approaches equation (1.5), that is, that of the hot-ball model, the sphere in a fluid with zero solute concentration, as expected.

Plots of values of $\ln(m/m_0)$, obtained from equation (1.39), are shown above in Figure 1.11. Curve 6 is for a value of hr tending towards infinity, that is, corresponds to the hot-ball model with zero concentration of solute in the fluid. Curves 1–5 are plots for values of hr of 1, 3, 6, 11 and 21, respectively. The time axis is expressed in units of t_c, where t_c is defined by equation (1.1), that is, the characteristic time for extraction with no solubility limitation. In all cases the curves become linear at longer times, as the terms for $n=1$ become dominant. At these times $\ln(m/m_0)$ is given approximately by

$$\ln(m/m_0) = \ln\{6(hr/a_1)^2/[hr(hr-1)+a_1^2]\} - (a_1/\pi)^2 t/t_c \tag{1.42}$$

The slopes of the linear portion of the graphs are thus equal to $-(a_1/\pi)^2/t_c$ and the intercepts $-I$ are given by

$$-I = \ln\{6(hr/a_1)^2/[hr(hr-1)+a_1^2]\} \tag{1.43}$$

Values of a_1 and I are given in Table 1.6 for the values of hr that were used in the plots of Figure 1.11. It can be seen that both these quantities fall as hr falls, that is, as the effect of solubility limitation increases. The effect of solubility limitation is thus to reduce the high rate at the beginning of the extraction, to reduce the slope of the linear portion, and to move the linear portion upwards on the graph. The extent of these effects increases as the pressure falls and the solubility decreases. The model is consistent with some of the experimental results described earlier.

Table 1.6 Parameters in the equations for the spherical model including solubility limitation and recovery after $0.3t_c$

hr	a_1	I	% recovery after $0.3t_c$
∞	3.1416	0.4997	63
21	2.9930	0.3731	57
11	2.8628	0.2884	52
6	2.6537	0.1887	46
3	2.2889	0.0866	36
1	1.5708	0.0146	23

The points shown in Figure 1.11 are the values of $\ln(m/m_0)$ at $t=0.3t_c$, with t_c defined by equation (1.1). The percentage recoveries, corresponding to these points on curves 1–5, are given in Table 1.6 for the various values of hr. In the limit of $hr=0$, the percentage extraction would also be zero, as zero solubility corresponds physically to no extraction. Although the relationship between percentage recovery and hr (and therefore solubility) is by no means linear, it shows that there is a good correlation between solubility and the percentage extracted at a particular time; this effect is most obvious at the beginning of the extraction. This result taken in conjunction with the fact that solubility in a supercritical fluid, in terms of concentration and at a temperature not too far above the critical temperature, is a strong function of pressure, means that the model predicts that a curve of percentage recovery versus pressure at short times is similar to a curve of solubility versus pressure, as shown in Figure 1.15. The model therefore predicts both the typical 'pressure-threshold' curves for extraction and the form of extraction kinetics.

1.5 Solubilities in supercritical fluids

It is important that extraction is carried out with a fluid and at a pressure so that the solute of interest is reasonably soluble. For quantitative analytical purposes, it is not important that the compound is extremely soluble, as extraction must be continued almost to exhaustion, by which time solubility is a less important factor. However, good solubility speeds up the initial stages of the extraction and reduces the time of the process to some extent. It may also be possible that conditions can be chosen so that the extraction of unwanted solutes that may interfere with the analysis is minimized. For these reasons, it is useful to have some knowledge of the solubilities of the solutes of interest and of those to be avoided. In this section, a few major points about these solubilities are made.

Firstly, it should be mentioned that solubilities are described either as the mole fraction of solute in the solution, x, or by the amount (as moles or mass) of solute per unit volume, that is, concentration, S, at saturation. The relationship between these quantities is trivial and involves the density of the solution in moles per unit volume, which, as the solutions are dilute, is often taken to be the density of the pure fluid. Thus, for example, S in moles per unit volume is given approximately by

$$S = x/V \tag{1.44}$$

where V is the molar volume of the pure fluid. Many publications have appeared in recent years giving experimentally obtained solubilities and those in pure CO_2, which appeared in Chemical Abstracts up to the end of 1989, have been reviewed [14]. To obtain values of V or its inverse, which is the density of the fluid, for use in equation (1.44) or in the density prediction methods described below, an equation of state for the fluid is required. Currently the most highly regarded equations of state for CO_2 are that of Ely [15] and the tabulated IUPAC formulation [16]. Equations of state adequate for analytical purposes are also sometimes included with SFE and SFC instrumental software.

The solubility of a substance in a supercritical fluid is contributed to by two factors, the volatility of the substance and the solvating effect of the supercritical fluid. The latter factor, as we shall see more quantitatively in the next

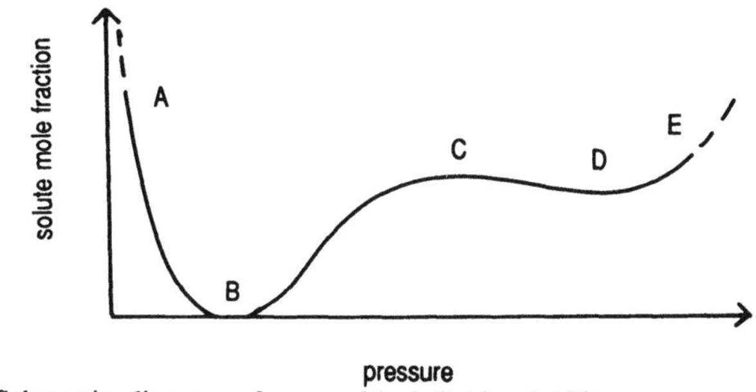

Figure 1.16 Schematic diagram of supercritical fluid solubility versus pressure at constant temperature.

section, is primarily a function of the density of the fluid. The solubility of a substance at constant temperature as a function of pressure and in terms of mole fraction has the schematic form of Figure 1.16. In section A–B, which is at very low pressures not of concern in most extraction processes, x falls as the solute is diluted by the fluid. In section B–C there is a rapid rise in x at a so-called threshold pressure characteristic of the solute–fluid system, which is a pressure somewhat above the critical pressure of the fluid. This occurs because of the rapid rise in the density, and therefore solvating effect, of the fluid at around this pressure. The next two features may or may not occur for any particular system, especially as the pressure range of interest may be limited. A fall, shown as C–D may occur because, at higher pressure, repulsive forces may 'squeeze' the solute out of solution. The rise shown as D–E occurs for moderately volatile solutes, if there is a critical line in the mixture phase diagram at higher pressures.

At constant pressure and as a function of temperature, solubility has a minimum at a particular temperature, as shown for phenanthrene in Figure 1.17. The initial fall occurs because, as the temperature rises, the density and solvating effect falls. However, as the temperature rises, the volatility of the solute also rises, and eventually this effect exceeds the effect of the falling solvation and the solubility rises.

The solubility of a substance is only a guide to its extractability. The solubility gives the concentration of solute in equilibrium with the pure solute. The presence of other materials in the matrix and in the fluid affect the equilibrium concentration. The effect of the matrix is, in general, to reduce this concentration below the solubility. If the matrix can be considered to be an ideal liquid mixture, the equilibrium concentration of each component is equal to its solubility multiplied by its mole fraction in the matrix. Most systems are non-ideal to an unknown extent, and the only conclusion that can be drawn is that the equilibrium concentration falls as the concentration in the matrix falls. The presence of other solutes in the fluid, in general, enhances the equilibrium concentration [17], as the solutes often have a stronger affinity to each other than to the solvent fluid.

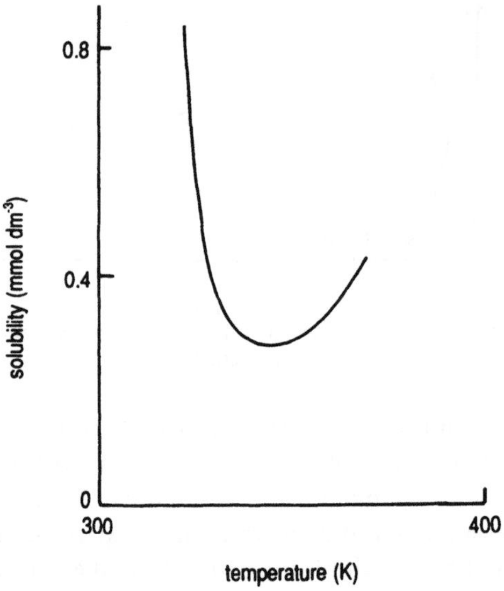

Figure 1.17 Solubility of phenanthrene in CO_2 at 100 bar as a function of temperature.

A severe similar effect, which is important in some cases, is that the extraction may involve replacement of solute molecules by fluid molecules on surface sites on the matrix. Although solubility, that is, the affinity of the fluid for the solute, is a factor in such a process, other factors like the Gibbs function for adsorption of both solute and solvent molecules are very important. These effects can sometimes be seen experimentally, when different supercritical fluids are found to extract different total amounts [18].

Nevertheless, in spite of these warnings, the extent of extraction of a solute after a given time, plotted as a function of pressure at constant temperature, often follows the same shape as the corresponding solubility curve and has approximately the same threshold pressure. As in the example given in Figure 1.15, the extraction behaviour follows the predicted solubility curve approximately.

1.5.1 Experimental methods for obtaining solubility in an analytical laboratory

It is perhaps unrealistic to expect that a preliminary study of the highest quality, as described in some of the solubility literature, would be carried out for an analytical application. The various methods described in the literature have been reviewed [14, 19]. However, an analytical laboratory involved in SFE and SFC will have apparatus on which studies of sufficient precision can be fairly easily and quickly made, and two types are described.

The first of these is a simplified version of one of the standard literature methods [20–26], in which an SFE apparatus of the analytical type is used to 'extract' the pure solute. Experimental extraction methods suitable for adaption for this type of measurement are described in the next chapter. Slow flow rates are used so that the effluent fluid is saturated, and this is checked by varying the flow rate and ensuring that the final result is unaffected. The extraction should be carried out at constant temperature and pressure. The flow rate at the supercritical fluid pump and temperature of the pump head should be measured and also the flow rate at a modifier pump, if one is used. The effluent is then collected for a given time and quantitatively analysed by any appropriate method, giving the number of moles of solute. The number of moles of solvent can then be obtained from the flow rate and density of the fluid at the pump (and also the same values for the modifier, if used) and the time of collection, and hence the mole fraction at saturation calculated. As an example, the solubility of anthracene, obtained by gas chromatographic analysis, following the trapping of a saturated solution [20] is shown in Figure 1.18.

A second method is to use chromatographic retention in SFC to produce qualitative 'pressure threshold' curves, which have approximately the same shape as solubility curves [27]. The degree of retention of a solute, as measured by the capacity factor, k', is at least qualitatively inversely related to the solvating power of the mobile phase for that solute: the more soluble it is in the mobile phase, the less it is retained. Thus the measurement of retention times at a given temperature and at a number of constant pressures, as well as the corresponding solvent retention times (which can be taken to be the mobile phase retention times) enables an isotherm of $1/k'$ versus pressure to be drawn.

This qualitative method can in some cases produce quantitative results. It can be shown experimentally that in some situations at constant temperature and for a particular column, the capacity factor for a particular solute is fairly accurately inversely proportional to its solubility in the mobile phase in terms

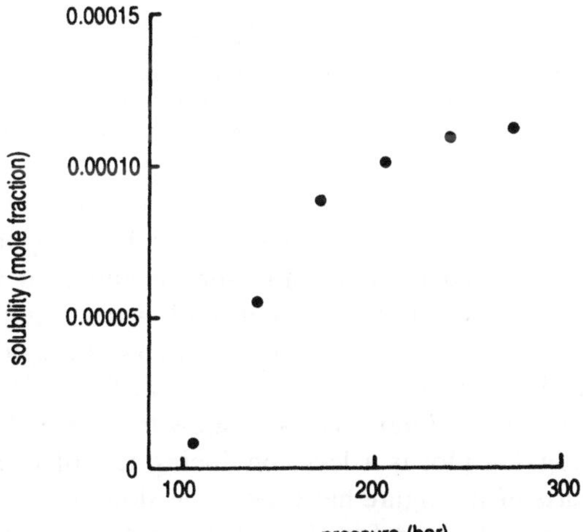

Figure 1.18 Solubility of anthracene in carbon dioxide at 45°C from the data of Kosal and Holder [20].

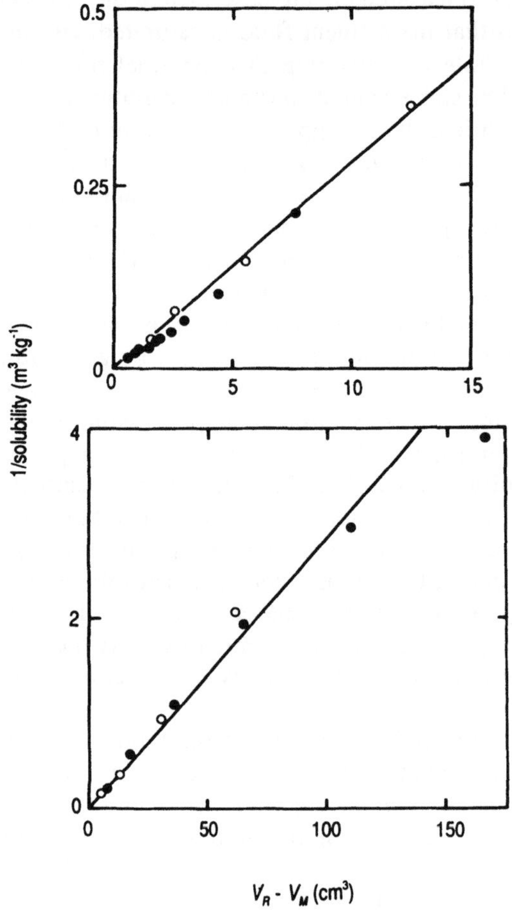

Figure 1.19 Relationship between solubility and retention volume (proportional to capacity factor) for naphthalene at 45°C using an ODS column. Open circles are methanol–water mixtures and closed circles supercritical CO_2. Lines on both graphs have the same slope [28].

of *S* [28]. Figure 1.19 shows 1/*S*, plotted against $V_R - V_M$ the retention volume of the solute minus that of the mobile phase (which is proportional to k') for naphthalene using a particular ODS bonded-phase column at 318 K. Although both CO_2 as a supercritical fluid and methanol–water liquid mobile phases were used, and the data cover a range of 2.5 orders of magnitude, the points all lie approximately on the same line. (Because of the wide range of values, two plots are given with different scales to show the low and high values; the alternative of a log–log plot is a less convincing test of linearity. The lines drawn on both parts of the figure have the same slope.)

Such plots are not obtained for all systems and it is necessary to check representative systems. However, the relationship can provide a relatively rapid way of generating large amounts of solubility data from $1/k'$, if the proportionality constant relating the two quantities is known. This constant may be obtained by a number of methods, including liquid solubility measurements or relatively few supercritical fluid measurements made by conventional methods. Solubility data, obtained in this way, have been published [29].

1.5.2 Prediction of solubility data

Prediction of supercritical fluid solubility data is difficult and not very accurate, even when there are some experimental values. However, some methods are available of sufficient reliability for the accuracy required to choose conditions for analytical SFE. The method best used depends on whether any experimental measurements have been made on the system of interest. Three methods are described.

Firstly, if a reasonable amount of data are available for a compound of interest and the other necessary physical parameters are also available, then an equation of state, such as the Peng–Robinson equation (P–R), may be used to correlate data over a limited range of pressure and temperature [30, 31]. This allows interpolation and limited extrapolation of the experimental data. An approximate version of the equations, which assumes dilute solutions and no effect of the fluid on the solute, is given in the first section of Table 1.7. The solute–fluid interaction parameter, δ, which is a fine-tuning parameter, is not known in advance and is obtained by fitting the equations to the experimental data, a procedure that also makes up for the approximations in the equations in the table [31]. The parameter, δ, is, in principle, temperature independent and so a value can be obtained by fitting all the data. Alternatively, isotherms can be fitted separately to the equations, in which case δ some-

Table 1.7 Simplified P–R equations for dilute solutions, giving the mol fraction, x, at saturation of a solute in a supercritical fluid (top section) with the parameters for solutions in CO_2 (bottom section) [32]

$$\ln x = \ln(p_v/p) - \ln\phi + pV_m/RT$$

$$\ln\phi = (b_2/b_1)(z-1) - \ln(z - pb_1/RT)$$
$$- (a_{11}/2\sqrt{2}\,b_1RT)(2a_{12}/a_{11} - b_2/b_1)\ln[(z + (1+\sqrt{2})\,b_1p/RT)/(z + (1-\sqrt{2})\,b_1p/RT)]$$

$$z = pV/RT, \quad b_1 = 0.0778RT_{c,1}/p_{c,1}, \quad b_2 = 0.0778RT_{c,2}/p_{c,2}$$

$$a_{12} = (0.45724(1-\delta)\,R^2T_{c,1}T_{c,2}\kappa_2\kappa_2)/\sqrt{p_{c,1}p_{c,2}}$$

$$a_{11} = (0.45724R^2T_{c,1}{}^2\kappa_1^2)/p_{c,1}$$

$$\kappa_1 = 1 + (1 - \sqrt{T/T_{c,1}})(0.37464 + 1.54226\omega_1 - 0.26992\omega_1^2)$$

$$\kappa_2 = 1 + (1 - \sqrt{T/T_{c,2}})(0.37464 + 1.54226\omega_2 - 0.26992\omega_2^2)$$

V is the molar volume of the pure solvent, p_v is the vapour pressure and V_m the molar volume of the pure solute. $p_{c,i}$, $T_{c,i}$ and ω_i are critical pressures, temperatures and acentric factors

Subscripts: 1 = solvent; 2 = solute

For CO_2: $p_{c,1} = 7.383$ MPa, $T_{c,1} = 304.2$ K, $\omega = 0.225$

$$\delta = 0.51A(\omega_2 - \omega_1)(V_{c,2}/V_{c,1})(p_{c,2}/p_{c,1})^2$$

$A = ½$ for compounds containing –OH groups and capable of forming hydrogen-bonded dimers

$A = 1$ for other compounds

$$V_{c,i} = (0.2918 - 0.928\omega_i)\,RT_{c,i}/p_{c,i}$$

times varies with temperature. This is especially the case when the solute is a solid that forms a liquid phase under certain conditions of temperature and pressure. P–R interaction parameters for 19 solutes in pure CO_2 have been published, along with the other physical parameters used [32].

Secondly, for a solute for which there are solubility data available, but not enough information on other physical parameters, or the will to carry out an equation-of-state correlation, less accurate estimates of solubilities may be made by the following, more empirical correlation. This is to fit the solubility in terms of mole fraction, x, at constant temperature as a function of pressure to the following equation:

$$\ln(xp/p_{\mathrm{ref}}) = A' + B(\rho - \rho_{\mathrm{ref}}) \tag{1.45}$$

where P is the pressure, ρ the density of the solution (approximately the density of the pure supercritical fluid), A' and B are constants at constant temperature and p_{ref} is a reference pressure, conveniently 1 bar. ρ_{ref} is a reference density, chosen to be central to the density range of all the solubility data. A good value for pure and modified CO_2 would be 700 kg m^{-3} and the following discussion assumes this value and also a p_{ref} of 1 bar. The reason for the choice of ρ_{ref} is to make the values of A' obtained much less sensitive to experimental error in the data and easier to correlate between sets of data at different temperatures. A full discussion of this correlation method has been given [14]. It is found that better correlations are obtained if data below 100 bar are not included in the fits and this has been done in the examples given below.

Values of A' and B are given in Table 1.8 for the solubility of naphthalene in CO_2, for which there is a large volume of data. The values of A' are also

Table 1.8 Values of the parameters A' and B obtained by fitting equation (1.45) to experimental solubilities for naphthalene in CO_2 [14]

T(K)	A'	$10^2 B$ (m^2 kg^{-1})
308	−0.1394	8.00
308	−0.1121	7.82
308	−0.1517	8.50
308	−0.2659	8.86
308	−0.1700	8.45
318	0.8542	6.50
318	0.8673	6.07
323	1.2267	6.70
328	1.7216	7.50
328	1.6802	7.84
328	1.6435	7.59
328	1.7143	7.91
328	1.7125	7.41
331.5	1.8086	8.02
333.4	2.3333	8.50
337.9	2.5105	8.77

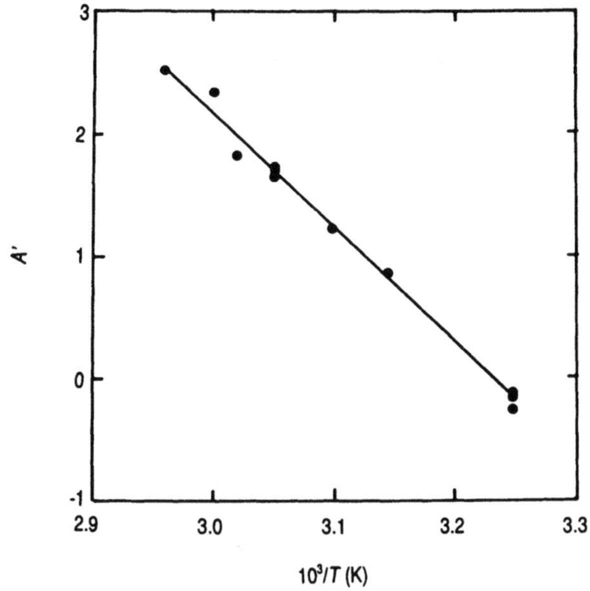

Figure 1.20 Plot of the parameter A' in equation (1.44) obtained from solubility isotherms for naphthalene in CO_2 versus $1/T$ [4].

shown plotted against $1/T$ in Figure 1.20. As can be seen, B does not vary greatly with temperature, whereas A' versus $1/T$ is approximately a straight line. The slope of the line is approximately $-H_v/R$, where H_v is the enthalpy of vaporization of the solute. In this case, the slope of the line corresponds to an enthalpy change of 77 kJ mol^{-1}, compared with a value for the enthalpy of vaporization of solid naphthalene of 70 kJ mol^{-1} [33].

The simplest prediction method using equation (1.45) is therefore to fit the experimental data at constant temperature to the equation and obtain values of A' and B at as many different temperatures as possible. It is then assumed that B is constant and if values at more than one temperature are obtained these are averaged, and the single or mean value used at all temperatures. For A' a value is obtained at any required temperature by plotting a graph similar to that of Figure 1.20. If there are enough experimental values to establish it, the best straight line is drawn through them. If not, a straight line of slope of $-H_v/R$ is drawn through (or close to) the point (or points). More sophisticated variants of this procedure have been described and values for the constants for the solubility of more than 60 compounds in CO_2 given [14].

Thirdly, for solutes for which no experimental data are available, approximate predictions can be made for solubilities in CO_2 using the P–R equation, and a correlation for the interaction parameter, δ [32]. This correlation is given in the lower section of Table 1.7, that is, all the equations of the table should be used in this method. Usually, estimates of many of the other necessary physical parameters and vapour pressures have to be made by standard methods [34]. Consequently, the solubility data obtained by this method are very approximate, although often sufficient for extraction optimization. The solubility data for atrazine, shown in Figure 1.15, were obtained in this way and based on scanty physical property data.

References

1. K. D. Bartle, L. Castle, A. A. Clifford, N. Din and M. Sharman, *in preparation.*
2. S. B. Hawthorne, *private communication.*
3. E. Brunner, W. Hultenschmidt and J. Schlichtharle, *J. Chem. Thermodynam.* **19** (1987) 273.
4. H. S. Carlslaw and J. C. Jaeger, *Conduction of Heat in Solids*, Clarendon, Oxford (1959).
5. J. Crank, *The Mathematics of Diffusion*, Clarendon, Oxford (1975).
6. K. D. Bartle, A. A. Clifford, S.B. Hawthorne, J.J. Langenfeld, D.J. Miller and R. Robinson, *J. Supercrit. Fluids* **3** (1990) 143.
7. H. J. V. Tyrrell and K.R. Harris, *Diffusion in Liquids*, Butterworth, London (1984).
8. K. D. Bartle, T. Boddington, A. A. Clifford, N. J. Cotton and C. J. Dowle, *Anal. Chem.* **63** (1991) 2371.
9. N. J. Mills, *Plastics — Microstructure, Properties and Applications*, Edward Arnold, London (1986).
10. K. P. Johnston, D.H. Ziger and C.P. Eckert, *Ind. Eng. Chem. Fundam.* **21** (1982) 191.
11. J. Kwiatkowski, Z. Lisicki and W. Majewski, *Ber. Bunsenges. Phys. Chem.* **88** (1984) 865.
12. K. D. Bartle, A. A. Clifford and N. J. Cotton, to be published.
13. S. Ashraf, K. D. Bartle, A. A. Clifford, M.W. Raynor and G. F. Shilstone, *Analyst*
14. K. D. Bartle, A. A. Clifford, S. A. Jafar and G. F. Shilstone, *J. Phys. Chem. Ref. Data* **20** (1991) K713.
15. J. F. Ely, *Proc. 65th Annual Convention Gas Processors Association*, San Antonia, Tx (1986) p.185.
16. S. Angus, B. Armstrong and K. M. de Reuck, *International Thermodynamic Tables of the Fluid State: Vol. III Carbon Dioxide*, Pergamon Press, Oxford (1976).
17. R. T. Kurnik and R. C. Reid, *Fluid Phase Equil.* **8** (1982) 93.
18. S. B. Hawthorne, *private communication.*
19. M. A. McHugh and V.J. Krukonis, *Supercritical Fluid Extraction — Principles and Practice*, Butterworth, Stoneham, (1986).
20. E. Kosal and G. D. Holder, *J. Chem. Eng. Data* **32** (1987) 148.
21. T. Bamberger, J. C. Erickson, C. L. Cooney and S.K. Kumar, *J. Chem. Eng. Data* **33** (1988) 327.
22. J. M. Dobbs, J.M. Wong and K. P. Johnston, *J. Chem. Eng. Data* **31** (1986) 303.
23. E. Stahl, W. Schilz, E. Schutz and E. Willing, *Angew. Chem., Int. Ed. Engl.* **17** (1978) 731.
24. E. Stahl and A. Glatz, *Fette, Seifen, Anstrichm.* **86** (1984) 346.
25. A. M. Pritchard, K.A. Peakall, E. F. Smart and G. J. Bignold, *Water Chem. Nucl. React. Systems* **4** (1986) 233.
26. S. K. Kumar, S. P. Chhabria, R. C. Reid and U. W. Suter, *Macromolecules* **20** (1987) 2550.
27. R. D. Smith, H. R. Usdeth, B. W. Wright and C. R. Yonker, *Sep. Sci. Tech.* **22** (1987) 1065.
28. K. D. Bartle, A. A. Clifford and S. A. Jafar, *J. Chem Soc., Faraday Trans.* **86** (1990) 855.
29. K. D. Bartle, A. A. Clifford and S. A. Jafar, *J. Chem Eng. Data* **35** (1990) 355.
30. D. Y. Peng and D. B. Robinson, *Ind. Eng. Chem. Fundam.* **15** (1976) 59.
31. K. D. Bartle, A. A. Clifford and G. F. Shilstone, *J. Supercrit. Fluids* **2** (1989) 30.
32. K. D. Bartle, T. Boddington, A. A. Clifford and G. F. Shilstone, *Proc. 2nd Int. Conf. on Supercritical Fluids*, Boston (1991).
33. *C R Handbook of Chemistry and Physics*, 52nd edition, CRC Press, Boca Raton, FL (1972) p. E43.
34. R. C. Reid, J. M. Prausnitz and B. E. Polling, *The Properties of Gases and Liquids*, McGraw-Hill, New York (1987).

2 Methodology for off-line supercritical fluid extraction

S.B. HAWTHORNE

2.1 Introduction

Supercritical fluid extraction (SFE) has long been investigated as an approach to extracting selected components from bulk samples (e.g. the classic extraction of caffeine from coffee) for a process engineering tool. In contrast, the application of SFE for the preparation of samples in the analytical laboratory has been receiving serious attention for only a few years. While process-scale SFE has generally been concerned with separating high concentration components from large quantities of a bulk matrix to increase their commercial value (or, e.g. to remove objectionable components as in the application of SFE to clean up contaminated soils), analytical-scale SFE is concerned with extracting analytes of interest from a bulk matrix as a sample preparation step prior to their determination by other analytical methods including (but not limited to) chromatographic, spectrometric, radiochemical and gravimetric techniques.

Much of the current interest in using analytical-scale SFE stems from the need to replace conventional liquid solvent extraction methods (e.g. extraction in a Soxhlet apparatus) with sample preparation methods that are faster, more efficient, have better potential for automation, and also reduce the need for large volumes of potentially hazardous liquid solvents. The need for alternative extraction methods is emphasized by current efforts by the US Environmental Protection Agency to reduce the use of methylene chloride as an extraction fluid for environmental sample preparation by 95% in the next few years [1]. The potential for applying SFE to a wide variety of environmental and biological samples for both qualitative and quantitative analyses has been discussed in recent reviews; the interested reader should refer to references [2–4] and the references therein. For reasons discussed later, analytical-scale SFE is most often applied to relatively small (e.g. several grams or less) samples, and methods to perform such extractions are the focus of this chapter.

Analytical-scale SFE can be broadly divided into two general approaches, off-line and on-line techniques. For the purposes of this chapter, off-line SFE refers to any method where the analytes are extracted using SFE and collected in a device independent of the chromatograph (or other measurement instrument). On-line SFE techniques utilize direct transfer of the extracted analytes to the analytical instrument (most frequently a chromatograph) to be used for the measurement of the extracted analytes. Throughout the remaining chapters of this book, on-line methods for combining analytical-scale SFE with chromatographic techniques including GC, SFC and HPLC are discussed. While the development of such on-line SFE methods of analysis has great potential for eventual automation and for enhancing method sensitivities [2–4], the great majority of analytical SFE results reported to date have utilized some form of off-line SFE followed by conventional chromatographic (or other method such as spectroscopic) analysis. The prevalence of reports using off-line SFE, particularly for quantitative work, is probably because off-line SFE is inherently simpler to perform.

To perform off-line SFE, only the SFE step must be successful and the extract can be analyzed at leisure by a variety of methods. The final product of an off-line SFE experiment typically consists of the extract dissolved in a few milliliters of liquid solvent, a form that is directly compatible with conventional chromatographic injectors. In contrast, the successful performance of on-line SFE requires that the SFE step, the coupling step (i.e. the transfer and collection of extracted analytes from the SFE to the chromatographic system), and the final chromatographic conditions be understood and controlled. With many on-line approaches, the sample extract is committed to a single analysis, and the extract is not available for analysis by other methods. While off-line SFE has the advantages of simplicity, availability of commercial instrumentation from several suppliers, and yields extracts that can be analyzed by several instrumental methods, on-line techniques have greater potential for enhanced sensitivity and automation as discussed in later chapters. Further development of on-line methods will doubtless increase their use for routine analyses, but it is likely that both off-line and on-line techniques will be employed for the foreseeable future. In any case, the majority of experimental and instrumental parameters discussed in this chapter for off-line SFE also relate to the SFE step of the various on-line techniques discussed in chapters 3–6.

2.2 The off-line SFE experiment

Off-line SFE is conceptually a simple experiment to perform and requires only relatively basic instrumentation as shown in Figure 2.1. The instrumental components necessary to perform SFE include a source of fluid (most often CO_2 or CO_2 with an organic modifier added to increase the fluid's polarity), a means of pressurizing the fluid, an extraction cell, a method of controlling the extraction cell temperature, a device to depressurize the supercritical fluid (flow restrictor), and a device for collecting the extracted analytes. In contrast to SFC, SFE is generally a constant pressure (not necessarily constant flow) experiment. Thus the pumping system needs to supply a constant pressure of

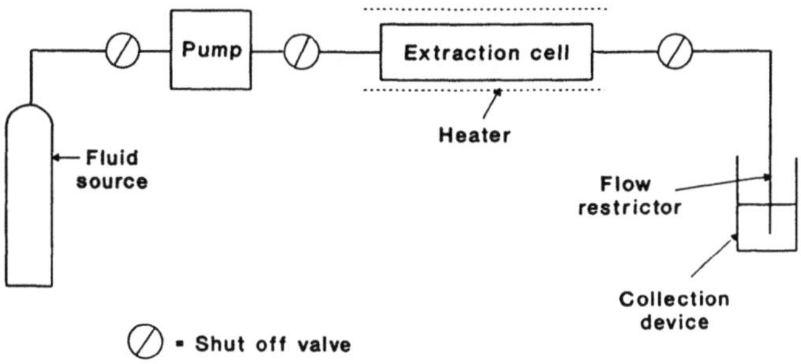

Figure 2.1 Basic instrumental components for performing off-line SFE. The shut off valve between the cell and the flow restrictor is required for static SFE, but not for dynamic SFE.

the extraction fluid, and the depressurization device (the flow restrictor) is normally used to control the flow rate of the supercritical fluid during SFE.

Two basic extraction modes, dynamic and static, have been widely used for analytical-scale SFE. In the dynamic mode, the sample is continually supplied with fresh supercritical fluid and the extracted analytes are constantly swept into the collection device. For static SFE, the outlet of the extraction cell is shut off and the cell is pressurized under static (non-flowing) conditions. Following an appropriate extraction time, the analytes are recovered from the static extraction, generally by opening a valve at the outlet of the cell and performing a short dynamic extraction. Since both static and dynamic modes are commonly used, and since they can affect instrumental requirements, the relative characteristics of these two modes are discussed throughout this chapter as appropriate.

The methods and devices used to complete each of the SFE tasks can greatly affect the ultimate extraction efficiencies, therefore the various designs and characteristics of each of these components are discussed in the following sections. Prior to a discussion of instrumentation, however, it is useful to divide the SFE experiment into its component parts.

For an SFE extraction to be successful, the target analytes must be controlled through three essential steps. First, the analytes must be partitioned from the sample matrix into the bulk supercritical extraction fluid. This step is controlled by the physicochemical interactions of the analytes, sample matrix, and supercritical fluid as described in chapter 1 by the 'extraction triangle' (i.e. analyte diffusion in the sample matrix, analyte solubility in the supercritical fluid, and analyte/matrix interactions). For the second (and easiest to understand) step during SFE, the analytes must be swept away from the matrix and out of the extraction cell. Third, the analytes must be efficiently collected from the supercritical fluid in a form that is compatible with the subsequent analytical method. Failure to achieve 100% efficiency for the target analytes during any of these steps results in less than quantitative recoveries.

Developing a quantitative SFE method is greatly helped by isolating each of these three steps (conceptually, and experimentally when possible) during the initial experiments. However, these steps are interdependent to some degree, and their relative effects on developing SFE methods are discussed in section 2.5. An awareness of these three steps is also useful for comparing various instrumental methods of performing off-line SFE. The first step, partitioning of the analytes from the matrix to the bulk supercritical fluid, is essentially a physicochemical problem (e.g. finding the SFE conditions that can best move target analytes from the matrix active sites into the bulk supercritical fluid), and an understanding of the partitioning processes helps to determine the relevant experimental requirements including pressure and temperature ranges and the need for pumping systems capable of introducing organic polarity modifiers.

Steps two and three are essentially 'plumbing' parameters. The supercritical fluid flows required to sweep the analytes from the extraction cell (step 2) helps to determine the pumping rate requirements, the maximum size of extraction cell that is useful, and the type of device (flow restrictor) used to control the supercritical fluid flow rate. The experimentally more difficult step 3, collecting the analytes after SFE, is nearly always performed by depressurizing the supercritical fluid through a flow restrictor and, therefore, helps to determine the characteristics of the flow restrictor and the analyte collection device.

Instrumentation for analytical-scale SFE has only recently become available from a number of suppliers, and there are several instrumental differences among manufacturers that can be discussed with consideration of the three steps that control SFE recoveries. While the relative infancy of analytical SFE almost guarantees that many instrumental changes will continue to occur, the following discussion describes presently available approaches to performing each part of the SFE experiment.

2.3 Instrumental considerations

2.3.1 Pumps

The majority of pumping systems used for SFE are similar to HPLC (or SFC) pumps based on either syringe or reciprocating piston designs, although alternative approaches such as a thermal pumping system have been reported [5]. Much early SFE work was performed with SFC pumps, or with HPLC pumps that had been modified to provide constant pressure (instead of constant flow). Recently, commercially available pumps specifically designed for SFE have become available from an increasing number of suppliers. It is not the intention of this work to recommend specific manufacturers, and any such recommendations would be futile since SFE instrumentation is rapidly developing. However, we have found several characteristics which are attractive for a pump used for analytical-scale SFE. (It should be noted that most fluids used for SFE, for example CO_2, are highly compressible, thus the control of the pump must be more sophisticated than for a liquid fluid to maintain the constant pressure condition.) The pressure rating of the pump should be at least 400 atm, and the pump should be capable of providing at least 2 ml/min of pressurized fluid at the upper pressure limit. The pump should recover to full pressure after pressurizing the extraction cell in a very few minutes (or in a few seconds if kinetic SFE investigations are to be performed as described below). Since syringe style pumps require refilling by the operator, such pumps should be capable of being filled and returned to the operating pressure in a few minutes (since fluids such as CO_2 are compressible, the pressurizing step can be much slower than pressurizing a liquid solvent). The ability to constantly read the pump flow rate is also valuable, particularly for determining total extraction fluid volume, and for extracting samples that are prone to restrictor plugging as discussed below. Finally, many SFE extractions require the addition of organic modifiers to supercritical CO_2 to obtain high extraction efficiencies, and the capability of dynamically mixing modifiers using a two pump system is especially attractive during the method development stage (although alternative methods are available as discussed below). The major cost of constructing an SFE instrument in the lab is the pumping system, and the pumps designed specifically for SFE can greatly increase the number of experiments that can be performed in a day. However, some alternatives to commercial SFE pumps exist, and can be particularly useful for initial investigations when funds are limited. For example, many HPLC syringe pumps do not have constant pressure capabilities, but can be 'tricked' into performing constant pressure SFE by setting the pressure limit switch to the desired pressure. Thus, the pump continues to cycle on and off around the set point pressure and provide fluid at an approximately constant pressure. HPLC pumps can also be utilized in the constant flow mode by utilizing a back-pressure regulator as shown in Figure 2.2 [6]. In this configuration, the flow

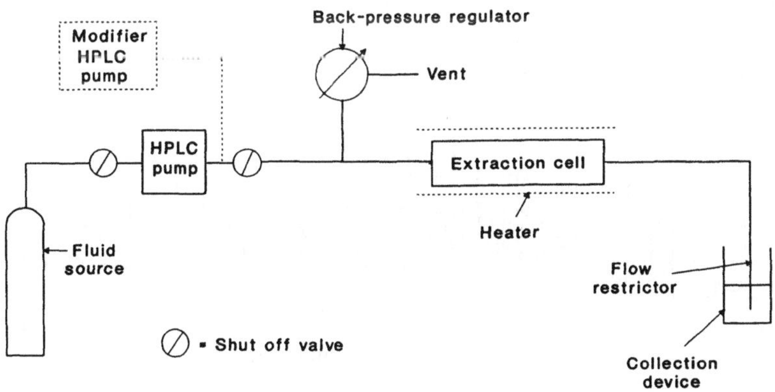

Figure 2.2 Schematic design for using a constant flow HPLC pump for constant pressure SFE. Note that the back pressure regulator must be heated to avoid plugging from frozen CO_2. A second flow-controlled pump may be added to perform dynamic modifier additions based on volume ratios.

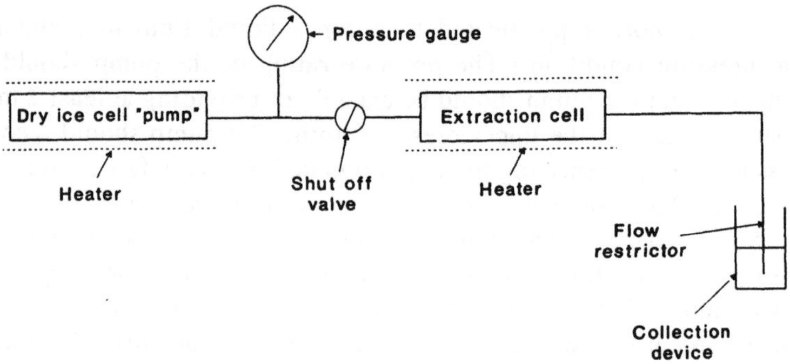

Figure 2.3 Simple SFE system using dry ice as the CO_2 source. For the sample that is shown in Figure 2.4, the dry ice was placed in the 'pump' cell and heated in a tube heater (a pipe wrapped with heat tape) to 100°C until the pressure gauge read approx. 170 atm. The shut off valve was then opened to extract the sample and the analytes were collected in approx. 2 ml of liquid solvent.

rate of the HPLC pump is set higher than the required SFE flow rate so that the extraction cell is pressurized until the set point pressure of the back-pressure regulator is exceeded. The regulator then opens to vent the excess fluid, thus maintaining a constant pressure supplied to the SFE cell. (Note that the back pressure regulator should be heated to avoid freezing of the CO_2 which is vented to maintain the constant pressure.) An even simpler (and cheaper) SFE pump can be constructed using two SFE cells as shown in Figure 2.3. The larger cell is filled with dry ice (solid CO_2) and heated to act as a pump to provide fluid to a second cell which contains the sample to be extracted. Surprisingly, a 10-ml cell that was filled with 9 g of crushed dry ice and then

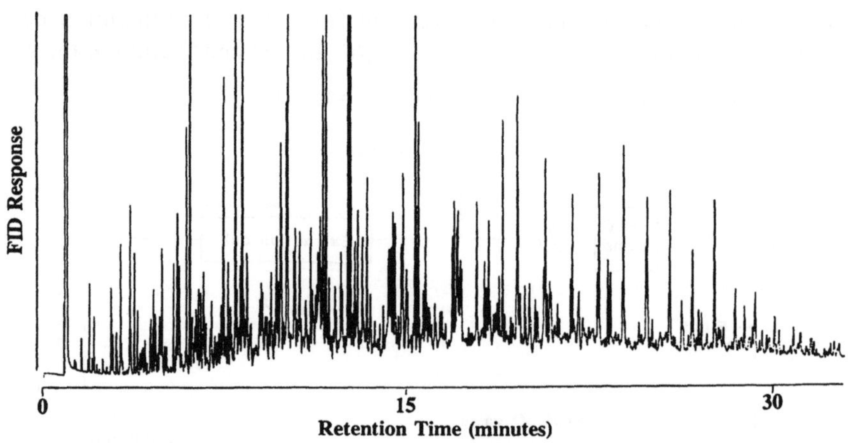

Figure 2.4 GC-FID chromatogram of an SFE extract of a coal tar contaminated soil using the simple dry ice pumping system from Figure 2.3. The sample was extracted for 30 min at about 170 atm. Major components were identified by GC-MS to be PAHs ranging in molecular weight from 178 to 228, and alkanes ranging from about C_{20} to C_{35}.

heated to 100°C was found to provide a constant pressure of approximately 170 atm CO_2 for over 30 min, and yielded good recovery of coal tar components from a contaminated soil as demonstrated in Figure 2.4. (The same sample was extracted for an additional 30 min using 400 atm and a conventional SFE system, and showed only small concentrations of organics suggesting that the dry ice pump extraction was approximately 95% efficient.) While this configuration is not likely to be useful for routine applications, it does demonstrate the inherent simplicity of off-line SFE.

2.3.2 Extraction cells

In the past, many investigators have utilized tube fittings (e.g. 'Swagelok') or HPLC columns, however, the recent availability of commercial SFE cells has reduced the need for 'home-made' units. Because of the need to hold relatively high pressures, the pressure rating and the absence of leaks are the most important characteristics of SFE cells. The ease in filling and handling the cell (e.g. finger-tight sealing versus the need for wrenches) should also be evaluated as well as the ability to replace individual cell components (e.g. seals and connectors). Although the early SFE cells were often subject to leaking, recent cells are much more reliable. We have had particularly good service using cells provided by JASCO (Hachioji, Japan) and Keystone Scientific (Bellefonte, PA) and literally hundreds of extractions have been performed using these cells without a leak, although care must be taken to ensure that no sample particles remain on the sealing surfaces when the cell is assembled. While the majority of SFE designs require that the cell withstand the extraction pressure, the need for high pressure materials can be eliminated using a cell design like that from the ISCO SFE instrument (Lincoln, NE). In this design, the sample cell is placed into a high pressure chamber, and both the inside and outside of the cell are pressurized at the same time. With such an approach, the need for relatively expensive stainless steel cells can potentially be eliminated. As shown in Figure 2.1, the temperature of the extraction cell must be controlled above the critical temperature of the extraction fluid, and this has generally been achieved either using an oven or a simple tube heater.

Some question exists concerning the best cell shape (i.e. long and narrow versus short and fat) for obtaining fast extractions. Long/narrow vessels are simpler to make and seal, while short/fat vessels are easier to fill. To determine the effect of cell shape on SFE recovery rates, we have extracted several samples under identical conditions (400 atm CO_2, 50°C, dynamic flow rate of ~0.6 ml/min) except that two different cells were used, a long/narrow (132 mm × 5 mm i.d.) and a short/fat (33 mm × 10 mm i.d.) cell, each cell having a total volume of 2.2 ml. Fractions of the SFE effluent were collected at specific time intervals and analyzed to determine the rate of recovery of PAHs from a 2.5-g sample of contaminated soil and flavor and fragrance compounds from a 1-g sample of lemon peel. As shown in Figure 2.5, the

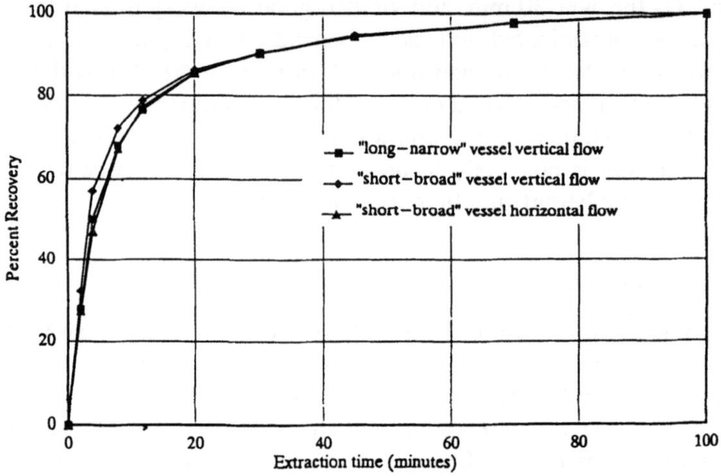

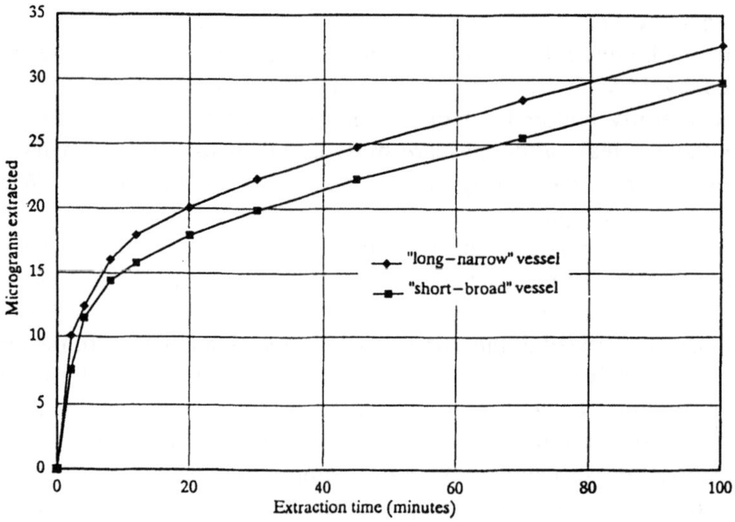

Figure 2.5 SFE recovery profiles of the PAH, fluoranthene, from contaminated soil (top) and a sesquiterpene from lemon peel using a long/narrow (132 × 5 mm) and a short/fat (33 × 10 mm) cell, each cell having a total volume of 2.2 ml. (Note that the difference in quantities of the sesquiterpene extracted from the lemon peel is a result of sample-to-sample variation, and does not indicate a different rate of extraction using the two cell shapes.)

extraction rates for representative analytes were virtually identical regardless of the cell shape and orientation (horizontal versus vertical), demonstrating that cost and convenience are more important characteristics of a cell than its shape (the small differences in the quantities of the sesquiterpene extracted from the lemon peel reflects differences in replicate samples, and not differences in extraction rates). Commercial SFE cells that are presently available for a reasonable price are generally limited to upper volumes of about 10–20 ml, and because of these limitations as well as the flow parameters discussed below, off-line analytical-scale SFE has primarily been applied to samples in the 1–20 g range.

2.3.3 *Flow control devices (restrictors)*

As discussed above, the SFE pump is responsible for providing a constant pressure of the extraction fluid, and the flow rate of the supercritical fluid is controlled by the flow restrictor placed at the outlet of the extraction cell. Two general types of restrictors have been used for the majority of off-line SFE studies. The first type, linear restrictors, are generally made by simply attaching a short (10–50 cm) piece of small inner diameter, fused silica tubing to the outlet of the extraction cell. Changing the inner diameter and length of the restrictor tubing allows the SFE flow rate to be controlled with relatively good accuracy. For example, when an extraction is performed with 400 atm CO_2 using 10-cm lengths of fused silica tubing as restrictors, the flow rates for 20-, 25- and 30-μm i.d. tubing are approximately 0.3, 0.6 and 1.0 ml/min, respectively. (Flow rates refer to the pressurized CO_2. The corresponding flow rate of the gaseous CO_2 at the outlet of the restrictors is about 500 times greater.) Since they are cut from bulk tubing, linear restrictors are very inexpensive, and can be routinely replaced at no significant cost. However, linear restrictors can be subject to plugging when samples are wet or contain very high concentrations of extractable material (discussed below).

At least two manufacturers (JASCO and Hewlett-Packard) use a variable orifice at the outlet of the cell to control the pressure of the fluid during the extraction. With this design, the pump provides a constant flow rate of the fluid, and the pressure is electronically controlled by the variable restrictor orifice. These variable restrictor pumping systems have the advantage of providing a constant flow rate of the supercritical fluid that is selected by the operator, but are much more expensive than linear restrictors (thousands of dollars versus a few cents). As discussed below, the present SFE units with variable restrictors also have fewer options for the analyte collection method.

While the variable restrictors avoid plugging by the ability to change the outlet flow, analysts using linear restrictors will inevitably encounter restrictor plugging during the extraction of some matrices. Several ways to plug a restrictor during SFE can be imagined, and it is useful to divide restrictor plugging into two categories, water plugging and plugging from other matrix components. Air-dried samples with low concentrations of extractable matrix components will rarely plug a linear restrictor, however, samples with significant water content will often plug the restrictor during the first few minutes of the extraction. This plugging occurs because the expanding supercritical

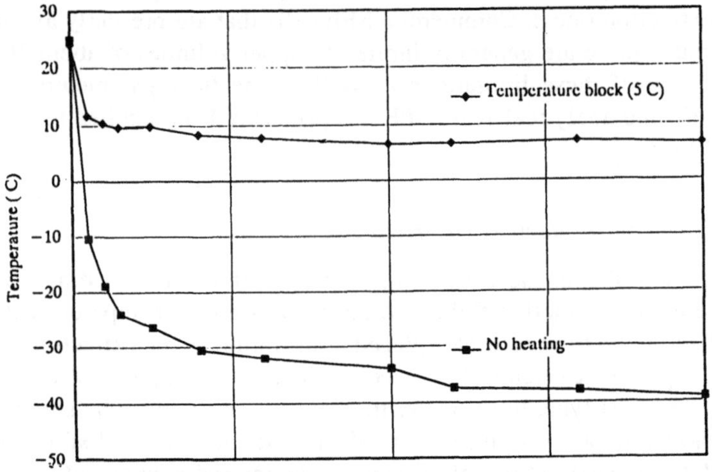

Figure 2.6 Collection solvent temperature (5 ml of methylene chloride) during off-line SFE using a supercritical CO_2 flow rate of approx. 0.6 ml/min (corresponding to approx. 300 ml/min gaseous CO_2). The bottom curve shows the solvent temperature with no solvent heating. The top curve shows the solvent temperature when the collection vial is placed into a heating block set at 5°C. The heating block was not pre-cooled, that is, the block was initially at room temperature.

fluid rapidly cools the restrictor tip (and the collection solvent) below the freezing point of water (Figure 2.6), thus causing the small amounts of water that are extracted to form an ice plug at the restrictor tip. This water plugging is easily solved by maintaining the collection solvent temperature above 0°C by placing the collection vial in a small beaker of water, or in a heating block that is set to turn on when the solvent temperature goes below 5°C [7].

Restrictor plugging from bulk matrix components is less frequent than water plugging, and tends to occur when the matrix has very high concentrations of extractable material (e.g. elemental sulfur from sediments, bulk hydrocarbons from petroleum waste sludges, fats from animal tissue). This type of plugging often occurs within a few seconds after starting the extraction, and is not avoided by simply maintaining the collection solvent temperature above 0°C. For such samples, the restrictor itself must be heated to avoid plugging, and restrictor heaters are available from manufacturers that use linear restrictors. However, the use of heated restrictors requires additional care to avoid the loss of more volatile analytes from the collection vessel as discussed below. Occasionally, the analyst may also encounter a sample that plugs the outlet frit of the extraction cell (which can easily be misinterpreted as restrictor plugging). We have experienced frit plugging only rarely, and the responsible samples were either sludges with very small (a few micrometers) particle sizes, or biological tissues that had little structural integrity (e.g. banana peel). In both cases, the samples could be extracted without plugging by simply placing a layer of sand about 1 cm deep between the sample and the cell outlet.

2.3.4 Analyte collection and recovery

Methods for recovering the analytes from the supercritical fluid have depended on depressurizing the fluid through the restrictor followed by collecting the analytes from the gaseous CO_2 stream. The development of efficient collection methods are not always straightforward, particularly for volatile analytes, simply because a relatively low flow rate of supercritical CO_2 yields a high flow of gaseous CO_2 (e.g. 1 ml/min depressurizes to about 500 ml/min). Three general approaches for collecting analytes off-line after depressurization of the extraction fluid are thermal trapping, sorbent trapping and trapping in a liquid solvent. Linear restrictors have been used with all three collection methods, however, present forms of the commercially available variable restrictors are difficult to couple with liquid solvent trapping, and have only utilized thermal and sorbent trapping.

Thermal trapping is inherently simple since the SFE effluent is simply depressurized into a cooled vessel. Unfortunately, quantitative thermal trapping has been limited to non-volatile organics since the high gas flow rate causes losses of even moderately volatile analytes. Non-volatile analytes may also be lost through the formation of aerosols. The two remaining approaches, trapping on a sorbent or in a liquid solvent have much better potential for quantitative recovery of high and moderate volatility analytes. Trapping on a sorbent is achieved by simply depressurizing the supercritical fluid onto the sorbent trap. Once the SFE is complete, the trapped analytes are recovered from the sorbent trap by eluting them with a small volume (a few milliliters) of liquid solvent. Sorbent trapping has the advantage of allowing the analyst to choose the sorbent packing that is best for the target analytes, as well as providing the possibility of selective collection on the trap, followed by additional selectivity during the elution of the analytes from the sorbent. For example, an SFE extract could contain three species with different polarities, A, B and C, of which only B is of interest. Species A could be removed by choosing a sorbent trap with low affinity for A, but high affinity for B and C, thus causing A to be lost through the trap during the SFE step. Species C could then be removed from the sorbent by first washing with a weak solvent, and finally, species B could be recovered in pure form by eluting the sorbent trap with a stronger solvent. While the possibilities for selectivity using sorbent traps are attractive, it should also be recognized that the sorbent trapping and subsequent elution add additional steps (which may or may not be quantitative) to the SFE experiment.

The simplest and most widely used method of trapping analytes involves simply depressurizing the supercritical fluid directly into a small vial containing a few milliliters of liquid solvent. This approach does not provide the

Table 2.1 SFE collection efficiencies of representative spikes in different organic solvents

	% Collection efficiency in 3 ml solvent[a]			
Test compound	Hexane	Methanol	CH_2Cl_2	CH_2Cl_2 (5°C)[b]
Phenol	43	55	77	98
2-Chloroethylether	60	56	73	100
1,2-Dichlorobenzene	46	58	78	100
Nitrobenzene	60	58	82	100
Isophorone	51	56	80	99
2-Nitrophenol	57	61	80	99
2-Nitroaniline	72	57	86	98
Acenaphthylene	66	60	84	99
Diethylphtalate	79	60	87	97
Hexachlorobenzene	79	62	86	100
Benzo[*a*]pyrene	76	57	92	97

[a] Collection efficiencies were determined by spiking 1-g samples of clean sand with approx. 18 μg of each component, and extracted using supercritical CO_2 at 400 atm (50°C) for 40 min at a fluid flow rate of approx. 0.6 ml/min (approx. 300 ml/min gas flow). Extracts were collected in a 33 mm × 12 mm i.d. glass vial containing 3 ml of the collection solvent. Each value is the average of three replicate extractions. Relative standard deviations were typically < 5%. Results are adapted from reference 7.

[b] Collection efficiencies when the collection vial was placed in a heating block set to turn on at 5°C. All other trials were performed by mildly heating the restrictor with a heat gun to mimic heating used to avoid restrictor plugging from frozen water.

potential for selectivity that sorbent trapping does, however, solvent trapping avoids the additional steps required by sorbent trapping. As is the case for sorbents, selection of the proper solvent polarity for the target analytes is important for achieving quantitative collection of the extracted analytes as shown in Table 2.1 by a comparison of the collection efficiencies obtained for several volatile and semi-volatile organics in different solvents. (Additional discussion of liquid solvent trapping conditions is given in section 2.5.) While not necessary for many analytes, setting the collection solvent vial in a temperature control block (e.g. set at 5°C) helps to avoid restrictor plugging from extracted water, while also increasing the collection efficiencies of volatile analytes by avoiding the temperature fluctuations caused by the use of other restrictor heating methods such as the use of a heat gun.

2.4 Supercritical fluids and modifiers

The characteristics of several fluids that have been used for SFE are given in Table 2.2, and criteria for selecting a supercritical fluid have recently been discussed [5]. Because of practical considerations, the majority of analytical-scale SFE investigations reported to date have used CO_2, primarily because it is relatively non-reactive, non-toxic, available at high purity, and has reasonable critical parameters. CO_2 is also capable of dissolving a wide range of non-polar and moderately polar molecules (solubilities of organics in super-

Table 2.2 Characteristics of representative supercritical fluids used for SFE

Fluid	P_C	T_C	δ[a]	Dipole moment (db)
CO_2	72.8	31.0	10.7	0.0
N_2O	71.5	36.4	10.6	0.2
NH_3	112.0	132.2	13.2	1.5
SF_6	37.1	45.6	7.6	0.0
CHF_3	48.0	26.1	8.7	1.6
$CHClF_2$	49.1	96.1	8.8	1.4
Ethane	48.2	32.2	8.7	0.0
n-Butane	37.5	152.0	7.7	0.0
Xenon	57.6	16.5	9.5	0.0

[a] Hildebrand solubility parameter in $(cal/cm^2)^{1/2}$. The values given are the maximum at high pressure.

critical CO_2 have recently been compiled by Bartle *et al.* [8]. Unfortunately, CO_2 is not a good solvent for polar analytes. In addition, some matrix/analyte interactions are sufficiently strong that, even though CO_2 is capable of dissolving an analyte, extraction recoveries are low because CO_2 does not compete well with the target analytes for active sites on the matrix. For some samples, including the extraction of PAHs from marine sediment [9] and the extraction of chlorinated dioxins from incinerator fly-ash [10], N_2O was superior to CO_2, even though they have very similar physical parameters (Table 2.2). Extraction efficiencies using $CHClF_2$ are even higher than those achieved using either CO_2 or N_2O as shown for PCB congeners from river sediment in Table 2.3. For this sample, however, N_2O had no apparent advantage over CO_2, while the 40-min SFE extraction with $CHClF_2$ yielded excellent recoveries compared to the 32 h of Soxhlet extraction.

Table 2.3 Comparison of Soxhlet extraction and different SFE fluids for the recovery of PCB congeners from river sediment

	PCB concentration (ng/g)[a]				
PCP congener	Soxhlet	$CHClF_2$	N_2O	CO_2	CO_2/MeOH
2,2′,3,5′	1070	1158	497	470	855
2,3′,4,4′,5 –	510	631	319	382	482
2,2′,3,4,4′,5,5′ –	160	166	145	144	168

[a] Each concentration was based on quadruplicate SFE extractions. Relative standard deviations were typically < 10%. The Soxhlet value is the concentration certified by the US National Institute of Standards and Technology based on two sequential 16 h Soxhlet extractions.

The routine use of N_2O as an SFE fluid should be discouraged since the combination of N_2O with easily oxidized organic material represents an explosion hazard. Despite its excellent characteristics as an SFE fluid, the routine use of $CHClF_2$ is also not likely because of the negative environmental effects of freons. Since other fluids that appear to have good properties for SFE generally suffer from being too reactive (e.g. ammonia), too expensive (e.g. xenon), or less polar than CO_2 (e.g. low molecular weight alkanes), the most practical approach is to increase the solvent strength of CO_2 by adding organic modifiers (generally in the range of 1–10% by volume). As shown in Table 2.3, CO_2 modified with 5% methanol yielded higher recoveries of the PCBs than pure CO_2 or pure N_2O. Similar increases in extraction efficiency using methanol modified CO_2 have been reported for a range of samples such as polar pesticides from soil [11], PAHs from environmental solids [9,12], ionic surfactants from soil and sludges [13], and chlorinated dioxins from soils [14]. However, methanol modified CO_2 does not always increase extraction efficiencies and different modifiers may often be required. For example, recent work in our laboratory showed no increase in the extraction efficiency of nitro-PAHs from diesel exhaust particulate matter when methanol modified CO_2 was compared to pure CO_2. However, toluene modified CO_2 yielded much higher recoveries (> 100% compared to Soxhlet extraction).

Although there is little in the literature to aid in selecting a modifier, a logical first choice for a modifier would be a liquid that is itself a good solvent for the target analytes. Specific interactions expected between the modifier and the sample matrix may also be useful to consider. However, as indicated by the extractions just discussed, choosing a modifier may rely heavily on trial and error, thus making rapid methods for surveying the effectiveness of several modifiers very attractive. There are several methods for adding modifiers to the CO_2, which vary considerably in cost, convenience and reliability. The simplest method is to simply pipet a small volume of the modifier into the extraction cell itself. While not very useful for dynamic SFE (since the modifier is rapidly purged from the cell), this method is particularly useful for static extractions since the concentration of the modifier remains constant during the static SFE step. This method also greatly simplifies the testing of many different modifiers in a short time.

A second method to introduce modifiers is to purchase premixed cylinders of modifier/CO_2. This approach is particularly good if the optimal composition of the modified CO_2 is known, and the analyst wishes to extract many samples under the same conditions. However, this approach is clumsy and expensive for selecting modifiers since a new cylinder must be purchased for each different modifier and concentration to be tested. A variant on the purchase of mixed modifiers is to simply fill an appropriate fraction of the (syringe) pump with modifier before filling with CO_2. While this method is mechanically simple and allows several modifiers to be evaluated in a reasonable time, the compressibility of CO_2 makes generating an accurate mixture difficult. Adding modifiers to the pumping system, either by hand or by purchasing mixed cylinders, runs the risk of contaminating pump components with the modifier itself.

Probably the best (and most expensive) method that is presently available for generating modified CO_2 is the use of dual pump systems that have recently become commercially available. One pump handles only pure CO_2, while the second pump provides a constant volume ratio of the organic modifier. With a dual pump system, the modifier concentration can be changed immediately, and the surveying of different modifiers only requires filling the second pump. Since the organic modifiers never come into contact with the CO_2 pump, the possibility of contaminating a pump is reduced. For laboratories that already have constant flow HPLC pumps, modifier mixing can also be accomplished using two pumps (set at the appropriate flow rates to achieve the desired volume concentration of the modifier) as shown in Figure 2.2. The total volume flow rate of the pumps is set above the outlet flow rate of the restrictor, and the constant pressure condition is maintained by the back pressure regulator [6].

With any method of adding organic modifiers to CO_2, it must be remembered that the critical temperature of the mixed solution is higher than that of pure CO_2 (32°C), therefore, the temperature of the extraction cell should be raised to ensure that a single phase supercritical fluid is present during the SFE. Although appropriate phase diagrams are available for CO_2/methanol [15], the data on critical temperatures for other CO_2/modifier combinations are generally not available. As a general rule, increasing the extraction temperature to 80°C for modifier concentrations in CO_2 up to 10% (v/v) should ensure that a supercritical condition is maintained for most organic modifiers.

Purity of the supercritical fluids and modifiers is also an important consideration, particularly when SFE is used to extract samples for trace analysis. Since the SFE experiment concentrates any system contaminants, all of the fluids and all of the SFE plumbing (including the pump, valves, cells, restrictors and all connecting tubing) must be very clean. High purity grades of CO_2 (SFE grades) have recently become available that have excellent purity characteristics, although our experience with the earlier (and less pure) SFC grades of CO_2 demonstrated sufficient purity for trace analyses using FID and MS detectors, while extracts analyzed using GC with an electron capture detector (ECD) generally showed significant background peaks. In general, however, we have found that the SFE instrumentation and associated plumbing is much more likely to cause background contamination than a good grade of CO_2. For this reason, it is very helpful to reduce the number of valves, transfer lines, and associated plumbing to a minimum in order to achieve clean blanks for trace work. All connections and fittings should be metal to metal and no Teflon tape or other fitting lubricants should be used. When contamination from the CO_2 (or other fluid source) is suspected, the gas cylinder should be confirmed as the contamination source by directly coupling the CO_2 cylinder to a restrictor (completely remove the SFE system) and purging CO_2 through a suitable collection solvent (or sorbent trap) using only the tank pressure. In our experience, this test will most often demonstrate that the contamination is in the SFE instrumentation, and not in the SFE or SFC grade CO_2.

High purity CO_2 is generally supplied with a dip tube (to allow liquid CO_2 to be transferred rather than gaseous CO_2) to facilitate rapid filling of the pump. Suppliers also provide CO_2 which has been pressurized with helium (e.g. at 1200 psi) which, in our experience, does not change the SFE results and can greatly aid pump filling.

2.5 Developing an off-line SFE method

Since analytical-scale SFE is a relatively new field of study, it is impossible to list a series of steps to follow that would guarantee the development of a successful SFE method. However, dividing the SFE experiment into its three component parts; partitioning of the analytes into the bulk supercritical fluid, sweeping the analytes out of the cell, and collecting the analytes from the depressurized CO_2, can greatly aid the methods development process. The partitioning step largely determines the extraction conditions including the choice of supercritical fluid (and modifier if used), extraction pressure and temperature, and the time required for the extraction. Sweeping the analytes out of the cell helps to determine the required SFE flow rates versus the sample size used. Finally, the conditions needed to quantitatively collect the analytes also help to determine the fluid flow rate as well as the configuration of the trapping device.

Before beginning this discussion, however, it is important to recognize that the use of spiked analytes to determine quantitative efficiency is frequently not valid for SFE (and for any other extraction method used for solid samples), simply because the native analytes are not necessarily in the same chemical or physical locations in or on the sample matrix. We have frequently observed that SFE conditions that yield quantitative recovery of spiked analytes fail to yield quantitative recovery of the native analytes. Thus, additional tests of quantitative recovery are necessary as discussed below. It should also be noted that the following discussion is designed only to describe some general ideas that we have found useful for developing an SFE method, and that the development of a successful SFE method will likely require interactive evaluation of conditions that optimize all three steps of the SFE experiment.

A logical first step is to determine the characteristics of the target analytes and choose initial SFE conditions. (If the target analytes are not known, preliminary extractions with pure CO_2, methanol-modified CO_2, or a liquid solvent can be performed and analyzed.) As previously discussed, CO_2 is the most practical fluid for SFE studies, although it does not have sufficient solvent strength to extract very polar analytes. If the target analytes are known to have significant solubility in supercritical CO_2 [8], pure CO_2 can be used for initial method development. For example, all of the analytes listed in Table 2.4 are sufficiently soluble to be extracted using supercritical CO_2, as long as matrix/analyte interactions are not too strong. If solubility data are not available, a general rule is that most compounds that can be analyzed using capillary GC have sufficient solubility in pure CO_2. As discussed below, however, high solubility does not guarantee that target analytes will be extracted from

Table 2.4 Estimated solubilities of representative organics in supercritical CO_2 at 400 atm and 50°C

Species	Solubility (mg/ml) at 400 atm, 50°C[a]
Naphthalene	190
Phenanthrene	13
Pyrene	3
Dibenzothiophene	11
Phenol	170
p-Chlorophenol	140
Docosane	320
Hexachloroethane	230
Diphenylamine	31
Palmitic acid	94

[a] Solubilities were estimated based on the tabulations given in reference 8.

the sample matrix. Since CO_2 is rarely too strong a solvent and since solvent strength of a supercritical fluid increases with pressure, it is reasonable to choose initial extraction conditions at the upper pressure limit of the available hardware (400 atm is a useful starting point). Other reasonable initial SFE conditions include a cell temperature of 40–50°C for pure CO_2 or 60–80°C for modified CO_2, a fluid flow rate of approximately 0.5–2 ml/min for 1- to 10-g samples, and an extraction time of 30 min.

After initial extraction conditions are selected, the efficiency of the two 'plumbing' steps should be evaluated. While spike recoveries are often not valid for determining whether the 'chemistry' part of the SFE has been solved, spike recoveries are an excellent means of evaluating the 'plumbing' steps. The entire focus of this spike recovery study is to ensure that the extracted analytes can be quantitatively swept from the cell and quantitatively collected under SFE conditions. Therefore, representative spikes are placed onto an inert matrix such as sand that would not be expected to strongly interact with the spikes. SFE is performed and the recovery of the spikes is determined. If the spike recoveries are not quantitative, additional extractions of the spiked matrix should be performed to determine whether the spikes were completely extracted. If no additional spikes are recovered, the analytes were most likely lost in the trapping step. For example, if volatile analytes are lost more than non-volatiles, poor trapping efficiencies are likely (or they were lost during the spiking process). When liquid solvent collection is used, the selection of solvent can be very important. As shown in Table 2.1, the collection efficiency of several organics varies greatly with the polarity of the collection solvent. Additionally, when the collection solvent was maintained at 5°C by placing the vial in a temperature control block (rather than heating the restrictor with a heat gun to avoid restrictor plugging from water), the collection efficiencies all approach 100% [7]. We have found that, for most analytes with vapor pressures less than heptane, collection by inserting the restrictor into approximately 3 ml of a suitable solvent yields >90% collection efficiencies.

However, changes in solvent composition, collection vial shape, flow rate, extraction time and restrictor temperature (when a restrictor heater is used) can all affect the trapping efficiencies, particularly of the more volatile components. Additionally, the use of such small volumes often requires that additional solvent be added to the vial during the SFE step to maintain sufficient solvent volume.

When sorbent traps are used to collect analytes from the supercritical fluid, less than quantitative recoveries can result either from failure of the sorbent to retain the analytes, or from failure of the rinse solvent to recover the analytes from the sorbent. A useful way to determine the efficiency of a sorbent trap is to simply place two sorbent traps in series. The presence of any analytes in the second trap indicates that the first trap is not quantitatively efficient. If no analytes are found in the first trap, either the spikes are not being extracted efficiently from the spiked matrix, or the solvent washing of the sorbent does not efficiently recover the analytes.

An antagonistic relationship between extraction flow rate and collection efficiencies can also exist, particularly for very volatile analytes. For example, when performing a 30 min extraction, the collection efficiency of α-pinene was 95%, 87% and 76% with extraction flow rates of 0.3, 0.6, and 1.2 ml/min, respectively (corresponding to gas flow rates of approx. 150, 300 and 600 ml/min, respectively). Thus, higher apparent recoveries sometimes result when a lower flow rate is used simply because the analytes are more efficiently trapped after the SFE step. The minimum useful flow rate can easily be determined for a sample by collecting and analyzing fractions during SFE. For example, when a 2.5 g sample of railroad-bed soil was extracted using flow rates ranging from 0.15 ml/min to 1.2 ml/min, the rates of recovery were essentially identical with flow rates of 0.3 ml/min and higher as shown in Figure 2.7 by the extraction kinetics for the PAH fluoranthene. The results in Figure 2.7 also demonstrate that attempts to state extraction conditions in terms of how many column volumes have passed through the extraction cell are not valid, as would be expected since the extraction of most native analytes is kinetically limited as described in chapter 1. In our experience, there is little or no advantage in extracting small samples (up to about 5 g) at an SFE flow rate of greater than approximately 0.5 ml/min, although larger samples may require somewhat faster flow rates to achieve optimal recovery rates. Because of these flow considerations and the lack of inexpensive extraction cells that can hold large samples, analytical-scale SFE is presently best applied to samples smaller than 15–20 g. For most real world samples, the SFE extraction rates do not appear to be limited by a lack of supercritical fluid (unless near saturation conditions exist, such as the extraction of fats from meats), but are limited by sample related kinetic factors as described below and in chapter 1. The understanding of the SFE rate limiting step is also important for deciding whether static or dynamic SFE modes should be used. When analytes are present at low concentrations (relative to their solubility in the SFE fluid),

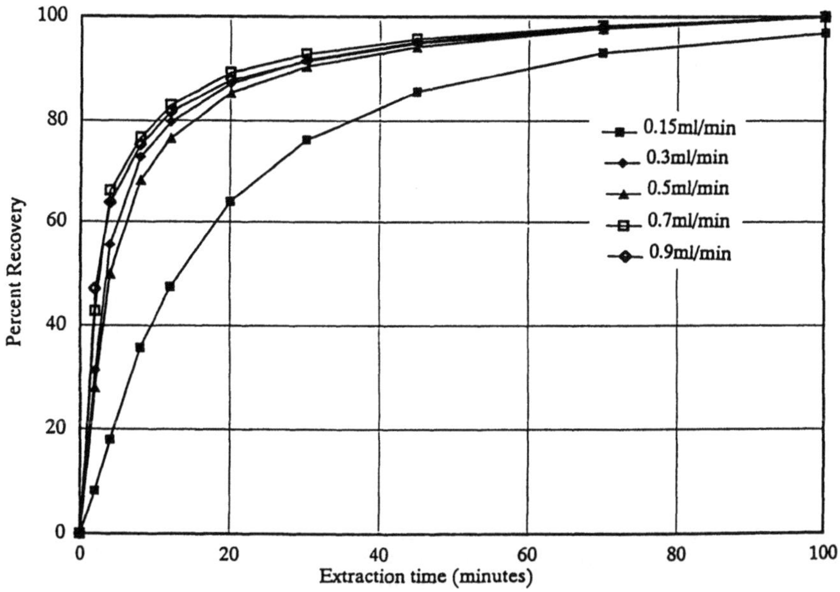

Figure 2.7 SFE recovery rates of the PAH fluoranthene from a 2.5 g sample of railroad-bed soil at different flow rates of supercritical CO_2.

dynamic SFE appears to have little advantage over static SFE. However, if the SFE recovery rates are limited by analyte solubility in the supercritical fluid, dynamic SFE (and higher extraction flow rates) has an obvious advantage over static SFE.

Once the spike recovery studies have demonstrated that the analytes can be efficiently swept from the cell and collected, evaluation of the real world samples should begin. As previously discussed, the partitioning of the analytes is controlled by the 'chemistry' of the system, that is, the factors that control the analyte partitioning between the matrix and the bulk supercritical fluid as discussed in chapter 1. This is by far the most difficult step to understand, since we generally do not know the physicochemical environment experienced by analytes in real world samples. If sufficient sample is available, it is generally useful to determine the extraction kinetics for individual analytes using the SFE conditions that yielded quantitative spike recoveries. Fractions are collected during the SFE experiment and analyzed, and the cumulative amount extracted is plotted versus extraction time. The reader is referred to chapter 1 for a theoretical discussion of extraction kinetics, however, observing the general shape of the extraction curve can direct further method development. Typical extraction kinetics curves are shown in Figure 2.8. Curve A shows typical recovery rates from a successful SFE extraction. If the curve rises more slowly, and continues to rise for an extended time (e.g. >45 min), the extraction is most likely limited by kinetics, either by diffusion in the matrix, or by slow desorbtion kinetics (Figure 2.8, curve B). If grinding the

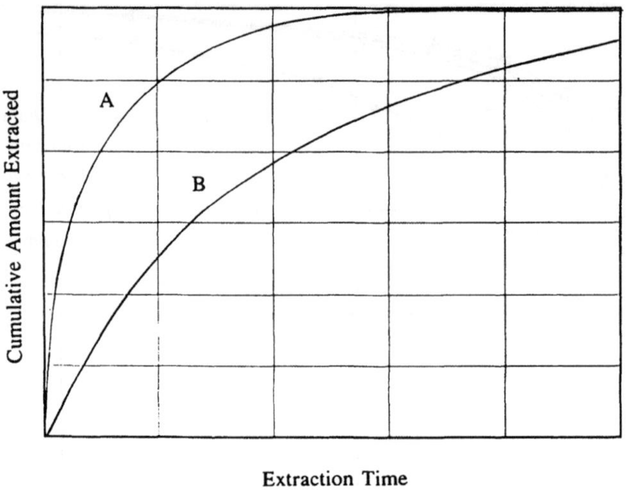

Figure 2.8 Typical SFE extraction kinetic curves for an analyte that shows no severe kinetic limitation (curve A) and for a species that shows a significant kinetic limitation to its SFE recovery (curve B).

sample yields faster extraction kinetics, then diffusion was a limiting factor. This is demonstrated in Figure 2.9 by the extraction of petroleum hydrocarbons from reservoir rock that was crushed and sieved to different particle sizes. However, if grinding the sample does not increase extraction rates, it is likely that the supercritical fluid does not efficiently compete with the analytes for matrix active sites, and that stronger SFE conditions are needed. This is

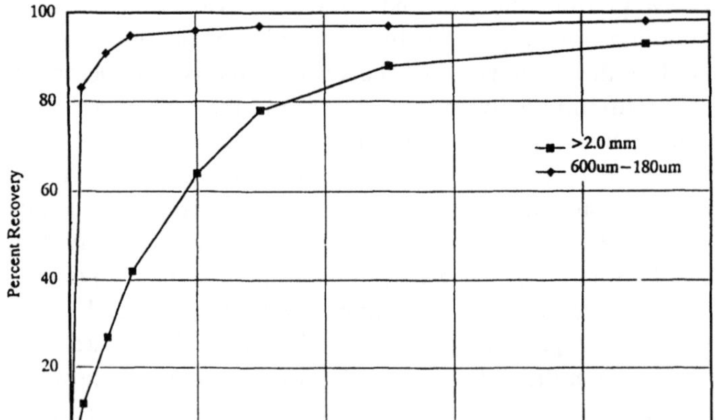

Figure 2.9 SFE recovery rates of *n*-pentacosane from petroleum source rock using particle sizes of 2–4 mm (bottom curve) and particle sizes of 180–600 µm (top curve). Both samples were extracted under identical conditions with 400 atm CO_2.

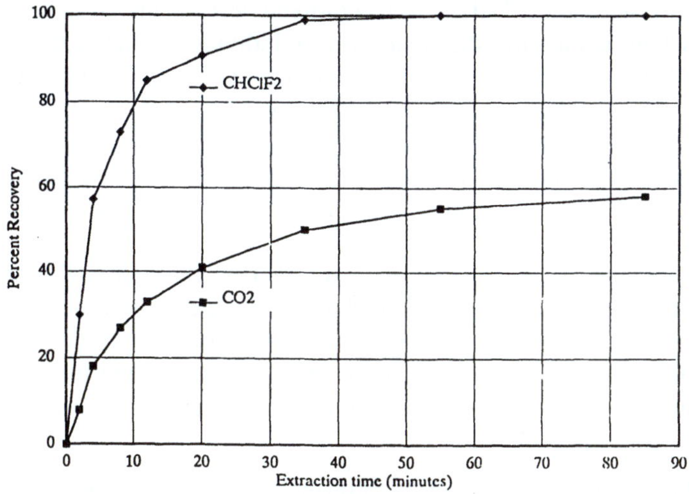

Figure 2.10 SFE recovery rates for the PAH pyrene from a wet petroleum waste sludge using supercritical $CHClF_2$ and supercritical CO_2 (both at 400 atm and approx. 0.6 ml/min, CO_2 at 50°C and $CHClF_2$ at 100°C).

demonstrated in Figure 2.10 by the extraction of pyrene from a wet petroleum waste sludge. For this case, supercritical CO_2 could not yield quantitative recovery of the native PAHs even after 90 min of extraction, even though they have more than sufficient solubility to extract (Table 2.4). Grinding the sample did not increase extraction rates, indicating that matrix/analyte interactions limited the extraction kinetics, and that a fluid that could better compete for matrix active sites was needed. However, when SFE was performed using pure $CHClF_2$, quantitative extractions of the PAHs were achieved in about 35 min as shown in Figure 2.10.

Since good spike recoveries do not necessarily indicate good recovery of native analytes (an unfortunate truth for any extraction method whether based on supercritical or liquid solvents), further means should be used to validate a proposed SFE method. Two very useful validation methods are to compare the SFE recoveries with conventional liquid solvent extraction methods on replicate samples, and to perform multiple sequential extractions (e.g. SFE followed by liquid solvent extraction) of a single sample. For example, the petroleum waste sludge shown in Figure 2.10 was extracted with methylene chloride (14 h sonication) after both the supercritical $CHClF_2$ and CO_2 extractions. As would be expected based on the results shown in Figure 2.10, the methylene chloride extract of the CO_2-extracted sludge contained substantial concentrations of PAHs. In contrast, the methylene chloride extract performed after the supercritical $CHClF_2$ extraction failed to yield any additional PAHs, indicating that supercritical $CHClF_2$ yielded quantitative recovery of the native PAHs (at least compared to the ability of methylene chloride to extract any additional PAHs). An additional test of the SFE method was performed by comparing the quantities of individual PAHs extracted from fresh sludge samples using methylene chloride (14 h sonication), and SFE for 40 min with both supercritical $CHClF_2$ and CO_2. As might be expected from Figure 2.10, SFE with CO_2 only yielded 50–80% recovery (versus methylene chloride) of the individual PAHs, while SFE with $CHClF_2$ actually yielded 20–40% more extracted PAHs than the 14 h methylene chloride extraction.

To summarize, a step-by-step approach that would lead to the successful development of quantitative SFE methods is not possible to describe at this time. However, a generalized approach can be based on the following considerations. For solving the 'chemistry' (analyte partitioning from the matrix to the bulk supercritical fluid) portion of the extraction problem the analyst should consider the following:

(1) If the target analytes have significant solubility in supercritical CO_2 (or can be analyzed using capillary GC), try initial extractions with 400–600 atm CO_2 at 40–50°C for 30 min at 0.5 to 1 ml/min flow rate. If the analytes are too polar to extract in pure CO_2, the addition of organic modifiers should be evaluated.
(2) Generate kinetic plots and/or perform multiple extractions (e.g. SFE followed by liquid solvent extraction) for real world samples. If these results indicate that SFE is not quantitative, grinding the sample will increase SFE rates if the recoveries are limited by diffusion in the matrix. If grinding the sample does not increase the extraction rates, add polarity modifiers and repeat the evaluation.

Concurrent method development to test the 'plumbing' aspects, that is, efficient sweeping of extracted analytes from the cell and quantitative collection of extracted analytes, should include the following:

(1) Determine collection efficiencies for spikes from a non-sorptive matrix. Develop collection method to ensure quantitative collection of extracted analytes under the SFE conditions used for quantitative extraction.
(2) If sample sizes are relatively large (e.g. > 5 g) determine the effect of supercritical fluid flow rate on recovery rates and efficiencies. This evaluation will also be useful to determine if static SFE can be used (which can greatly simplify the addition of polarity modifiers).

2.6 New directions in off-line analytical SFE

The majority of analytical SFE studies reported to date have focussed on the extraction of relatively non-polar organics (e.g. species that can be separated using GC) from solid matrices, and this chapter has dealt primarily with such

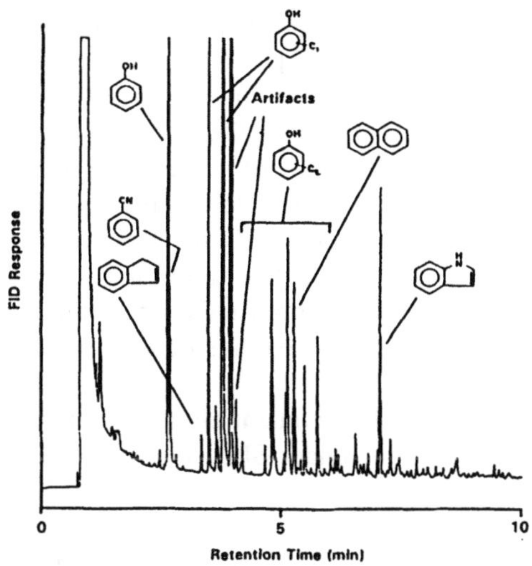

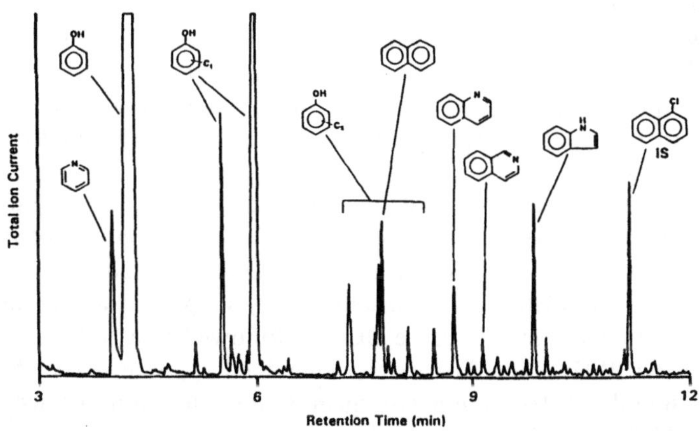

Figure 2.11 GC-MS analysis of an SFE extract from a 4 ml sample of gasification wastewater (bottom) and GC-FID analysis of an 'Empore' C18 sorbent disc used to collect the organics from a 100 ml sample of the same wastewater (top). The water was adjusted to pH 2 before passing through the sorbent disc to aid in the collection of the phenolic contaminants.

extractions. However, very encouraging results have been reported for the extraction of polar, higher molecular weight and ionic species using modified CO_2, including SFE of polar pesticides and ionic surfactants [4,11,13], or using different pure fluids such as $CHClF_2$ for SFE of steroids [16]. Recent reports have also described potentially valuable variations on the SFE experiment including the direct extraction of water samples [17], class-selective SFE by carefully selected extraction conditions [18], and the use of *in situ* chemical derivatization during SFE to reduce the polarity (and increase the extraction efficiency) of ionic analytes [19–21].

Direct SFE of water has been performed by inverting the flow of the supercritical CO_2 (so the CO_2 enters from the bottom of the extraction cell and exits from the top [17]). An example of using direct SFE for the extraction of a coal gasification waste water is shown in Figure 2.11 (bottom). An attractive alternative approach to using SFE for the preparation of water samples is to first run the water through a solid phase adsorbent (e.g. the C-18 'Empore' sorbent discs available from Analytichem), then recover the analytes using SFE as shown for the same coal gasification wastewater in Figure 2.11 (top).

The use of *in situ* chemical derivatization has recently been reported for the derivatization and extraction (under SFE conditions) of bacterial phospholipids (as their fatty acid methyl esters) from bacteria [20, 21], silylation and extraction of coffee components [19], and acid herbicides (as methyl esters) [20]. The derivatization of analytes in the sample matrix may have a particular advantage for SFE extractions since the less polar derivatives are easier to extract, and are easier to analyze using conventional GC techniques. Derivatizing reagents could also be selected to increase detector sensitivity and selectivity. For example, we have recently used chemical derivatization under SFE conditions to form the pentafluorobenzyl ester of the herbicide 2,4-dichlorophenoxyacetic acid (2,4-D), which both makes the herbicide amenable to GC analysis and greatly increases its sensitivity for electron capture detection.

Class-selective extractions using SFE have the potential to yield extracts free of matrix components that may interfere with the analysis of the target analytes. Class-selective extractions of alkanes (C20 to C30) and PAHs from diesel exhaust particulate have been achieved by simply extracting the sample first with 75 atm CO_2 (to extract the alkanes) followed by a 300 atm extraction (to extract the PAHs). The first extract contained approximately 85% of the alkanes, while the PAHs were nearly quantitatively found in the second fraction [22]. Unfortunately, the use of two SFE conditions to achieve class-selective extractions appears to be highly matrix dependent. For example, when we used the same approach to extract a petroleum waste sludge (which contained a similar distribution of alkanes and PAHs), virtually no selectivity was achieved even though the target analytes were the same as for the diesel exhaust particulate. Class-selective SFE of analytes from intermediate sorbents has been successful for the fractionation of alkanes, alkenes and aromatics from gasoline [23] and of saturates, aromatics and asphaltenes from crude oil [24]. Silica columns have also been used to enhance the selective extraction of low ppm concentrations of pesticides from lard using CO_2 with 2% methanol modifier. Excellent (>95%) recoveries of several organochlorine pesticides were obtained, yet the extracts were sufficiently fat-free to be analyzed directly by capillary GC [18].

2.7 Summary

Off-line SFE can be used to quantitatively extract a wide range of analytes from a variety of matrices. Although SFE instrumentation is still under development, suitable instruments can easily be assembled in the lab, and several vendors now offer useful components and complete systems. To develop quantitative methods for SFE, it is important to ensure that the analytes are efficiently extracted from the sample matrix, are efficiently swept from the extraction cell, and efficiently collected after depressurization of the supercritical fluid. While SFE has primarily been applied to relatively non-polar analytes, it is likely that significant advances in the extraction of polar analytes and analytes that are strongly associated with sample matrices will be achieved as the action and use of polarity modifiers is better understood. Recent reports describing SFE of water samples, class-selective SFE, and chemical derivatization under SFE conditions demonstrate the potential for using SFE techniques to replace many of the sample preparation steps that have traditionally been performed using liquid solvents.

Acknowledgements

The authors would like to thank the US Environmental Protection Agency, the New Jersey Department of Environmental Protection (Office of Science and Research), Shell Oil Company (USA), British Petroleum (USA), Procter and Gamble Company (USA), and the US Department of Energy for support of their SFE investigations. Instrumental support from ISCO (USA) is also gratefully acknowledged.

References

1. W. Budde, US Environmental Protection Agency, Cincinnati, OH, personal communication.
2. J. L. Veuthey, M. Caude and R. Rosset, *Analysis* **18** (1990) 103.
3. R. W. Vannoort, J. P. Chervet, H. Lingeman, G. J. DeJong and U. A. Th. Brinkman, *J. Chromatogr.* **505** (1990) 45.
4. S. B. Hawthorne, *Anal. Chem.* **62** (1990) 633A.
5. J. Pawliszyn, *J. High Resolut. Chromatogr.* **13** (1990) 199.
6. K. Anton, N. Pericles and H.M. Widmer, *Proc. 10th Int. Symp. on Capillary Chromatography*, Riva del Garda, Italy (1989) p. 1325
7. J. J. Langenfeld, M. D. Burford, S. B. Hawthorne and D. J. Miller, *J. Chromatogr.* in press.
8. K. D. Bartle, A. A. Clifford, S. A. Jafar and G. F. Shilstone, *J. Phys. Chem. Ref. Data* in press.
9. S. B. Hawthorne, D. J. Miller and J. J. Langenfeld, *J. Chromatogr. Sci.* **28** (1990) 2.
10. N. Alexandrou and J. Pawliszyn, *Anal. Chem.* **61** (1989) 2770.
11. J. R. Wheeler and M. E. McNally, *J. Chromatogr. Sci.* **27** (1989) 534.
12. S. B. Hawthorne and D. J. Miller, *Anal. Chem.* **59** (1987) 1705.
13. S. B. Hawthorne, D. J. Miller, D. D. Walker, D. E. Whittington and B. L. Moore, *J. Chromatogr.* **541** (1991) 185.
14. F. I. Onuska and K. A. Terry, *J. High Resolut. Chromatogr.* **12** (1989) 357.

15. E. Brunner, W. Hultenschmidt and G. Schlichthrale, *J. Chem. Thermodyn.* **19** (1989) 273.
16. S. F. Y. Li, C. P. Ong, M. L. Lee and H. K. Lee, *J. Chromatogr.* **515** (1990) 515.
17. J. L. Hedrick and L. T. Taylor, *J. High Resolut. Chromatogr.* **13** (1990) 312.
18. J. E. France and J. W. King, *J. Agric. Food Chem.* **39** (1991) 1874.
19. J. W. Hills and H. H. Hill, *Proc. Int. Symp. on Supercritical Fluid Chromatography and Extraction*, Park City, UT (1991) 113.
20. D. J. Miller, S. B. Hawthorne and J. J. Langenfeld, *Proc. Int. Symp. on Supercritical Fluid Chromatography and Extraction*, Park City, UT (1991) p. 155.
21. D. C. White, D. E. Nivens, D. B. Ringelberg and D. B. Hedrick, *Proc. Int. Symp. on Supercritical Fluid Chromatography and Extraction*, Park City, UT (1991) p. 43.
22. S. B. Hawthorne and D. J. Miller, *J. Chromatogr. Sci.* **24** (1986) 258.
23. J. M. Levy and J. P. Guzowski, *Fresenius Z. Anal. Chem.* **330** (1989) 207.
24. H. Skaar, H. R. Norli, E. Lundanes and T. Greibrokk, *J. Microcol. Sep.* **2** (1990) 222.

3 Coupled (on-line) supercritical fluid extraction-gas chromatography

S.B. HAWTHORNE

3.1 Introduction

The development of techniques to directly couple sample extraction/preparation steps with chromatographic analysis has received significant attention because of the potential to achieve very rapid, sensitive and cost-effective analyses. In general, the development of coupled or on-line extraction/analysis methods has been most successful when the extraction fluid is of the same phase as the chromatographic carrier fluid. For example, the combination of headspace analysis (or thermal desorption) with capillary gas chromatography (GC) has been widely used on a routine basis, and on-line instruments have been commercially available for several years. While gas-phase extraction methods such as headspace analysis are easily coupled with gas chromatography, the range of potential analytes is severely limited by the requirement of high analyte volatility. Therefore, nearly all extractions of moderate to low volatility analytes from solid samples have required liquid solvent extraction before GC analysis. Unfortunately, on-line methods for the coupling of liquid solvent extraction with capillary GC are few, primarily because the large volumes of liquid solvent used for the extraction step are incompatible with capillary GC columns. As a result, only a tiny fraction (e.g. 1 μl) of a liquid solvent extract is usually introduced into a GC column for the vast majority of liquid solvent extracts.

An ideal on-line extraction/GC method would combine the ability of liquid solvent extraction to efficiently extract a broad range of analytes with the ability of gas-phase extraction methods (e.g. headspace analysis) to rapidly and efficiently transfer the extracted analytes to the gas chromatograph. Within certain limitations to be discussed in this chapter, the characteristics of supercritical fluids make them ideal for the development of on-line sample extraction/gas chromatographic (SFE-GC) techniques. As discussed in chapters 1 and 2, SFE has the ability to extract many analytes from a variety of matrices with recoveries that rival liquid solvent extraction, but in much shorter extraction times. Additionally, since most supercritical fluids are converted to the gas phase upon depressurization to ambient conditions, SFE has the potential to introduce extracted analytes to the GC in the gas phase, without the problems related to introducing large volumes of liquid solvents.

The reader who is interested in the variety of qualitative and quantitative applications that have been explored using SFE-GC is referred to recent reviews and the references therein [1, 2]. This chapter focusses on the development of, and methods to perform on-line SFE-GC. While SFE-GC is a new approach to coupled sample extraction/chromatographic methods and new investigations are certain to improve present approaches, an attempt is made to discuss general factors to be considered before implementing an SFE-GC method including simplicity, reliability, and analytical sensitivity and reproducibility. Since the vast majority of GC separations utilize capillary (e.g. 250–320 μm i.d.) columns, the discussions in this chapter emphasize the coupling of SFE with capillary GC.

3.2 Generalized approaches to SFE-GC coupling

The successful performance of coupled (or on-line) SFE-GC requires four essential steps. Although these steps may appear obvious, an understanding of each step (and an understanding of how the target analytes behave during each step) is necessary to implement a method that can yield reliable qualitative and quantitative results. First, the analytes must be efficiently extracted from the sample matrix, and the considerations for achieving quantitative extraction of the target analytes that have been discussed in chapters 1 and 2 apply. Second, the analytes must be transferred from the SFE system to the GC. Third, the supercritical fluid must be depressurized to the gaseous state and removed from the chromatographic system. (Note: The order of steps two and three can be reversed as described below). Fourth, since a typical SFE extraction requires several minutes, and since a typical capillary GC peak is only about 1 s wide, the extracted analytes must be refocussed in a narrow band to obtain good chromatographic peak shapes and efficient separations. After these essential steps are completed, the GC separation is performed in a normal manner.

Since the first report of coupled (or on-line) SFE-GC in 1986 [3], a variety of approaches has been used to couple SFE extraction with capillary GC chromatography [4–15]. These approaches can be loosely categorized into three groups as shown schematically in Figure 3.1. The first approach shown in Figure 3.1 (top) utilizes an external loop through which the SFE extract flows [4,5]. To analyze the SFE extract, the sample loop valve is rotated so that a 'heart cut' of the extract is introduced into the GC. In this configuration, the supercritical fluid is depressurized when the valve is rotated to deliver the extracted analytes in the gas phase into the GC. The gaseous analytes are transferred into the GC by sweeping the heated transfer line and sample loop with the GC carrier gas. The sample loop valve should also be heated to avoid deposition of the analytes in the sample loop when the extraction fluid is

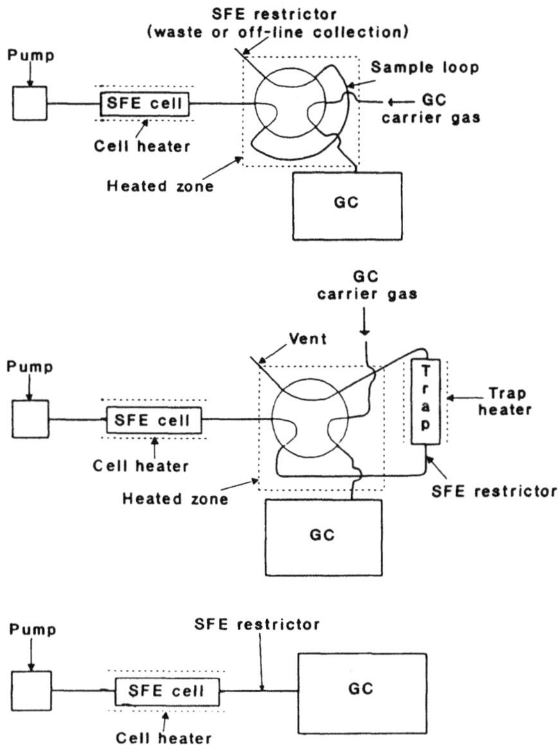

Figure 3.1 Schematics of three general approaches to SFE-GC coupling. The top approach utilizes a sample loop and valve to introduce a 'heart cut' of the flowing SFE extract to the GC through a heated transfer line by rotating the valve counterclockwise from the position shown. The middle approach utilizes a cryogenic or sorbent trap external to the GC to collect the SFE analytes from the depressurized fluid. The analytes are then transferred to the GC by rotating the valve counterclockwise, heating the trap, and sweeping the analytes into the GC. Both the top and middle approaches require that the valve and transfer lines are adequately heated to ensure that the analytes are efficiently swept into the GC with the GC carrier gas. With 'direct injection' SFE-GC (bottom), the analytes are directly transferred to the GC from the SFE cell by the extraction fluid without the need for intermediate valves or traps.

depressurized as the valve is rotated into the transfer line. Once the analytes are transferred to the GC system, they generally must be refocussed to obtain good chromatographic peak shapes (discussed later in the text).

A disadvantage of this approach is that only a fraction of the extracted analytes are transferred to the GC, however, the external loop approach does allow off-line collection and monitoring of SFE effluents, and select fractions of the SFE extract can be transferred to the GC for analysis. This concept has been successfully applied to class-selective analyses such as the selective determination of alkane, alkene and aromatic components of gasoline components after SFE extraction of the sample from silica columns [6].

In contrast to the external loop approach just discussed, the other two general approaches to SFE-GC are designed to transfer every extracted analyte molecule into the GC, and thus have the potential to yield maximum sensitivity. The first of these approaches utilizes a trap external to the GC to recover the analytes from the depressurized supercritical fluid as shown in Figure 3.1 (middle) [7–9]. Trapping of the analytes has been achieved using sorbent traps or a cryogenic loop (e.g. a cooled coil of stainless steel or fused silica tubing). Note that the cryogenic trapping temperature must be above the sublimation or boiling point of the extraction fluid. For example, CO_2 extracts cannot be trapped at liquid nitrogen temperatures since the CO_2 will also freeze and plug the collection device. After the SFE extraction is completed, the analytes are introduced into the GC by heating the trap and sweeping the trap with the GC carrier gas or with the gaseous extraction fluid. As was the case for the loop method described above, the analytes generally must be refocussed in the GC column to obtain good chromatographic peak shapes, particularly for more volatile analytes (a discussion of refocussing techniques in the GC column is given below).

Perhaps the simplest approaches to SFE-GC involve direct use of conventional GC injection ports (or a similar heating device) to achieve coupling with the SFE system [3,7,10–15]. These 'direct injection' approaches (Figure 3.1, bottom) depend upon collecting and refocussing the extracted analytes in the GC itself, either in the GC column's stationary phase or in a retention gap placed at the head of the analytical column. Such methods can be relatively simple to implement in a conventional GC laboratory since they do not require any significant alteration of conventional gas chromatographic instrumentation (i.e. the same GC can easily be used for both conventional liquid solvent injections and for SFE-GC studies with little or no hardware changes). Additionally, these simple approaches are responsible for the majority of qualitative and quantitative SFE-GC results that have been reported to date. For these reasons, and because an understanding of these relatively simple approaches should facilitate the development and implementation of more sophisticated SFE-GC techniques, the remainder of this chapter focusses on simple and direct SFE-GC coupling methods using conventional GC injection ports. Rather than provide an in-depth discussion and evaluation of every approach to SFE-GC that has been published, the goal of the remaining portion of this chapter is to provide the reader with a suitable background to construct and operate a simple on-line SFE-GC instrument. The hope is that the reader, by constructing and using such a system, will understand the principles of SFE-GC and obtain relevant experience with a minimum of time, expense and disruption of their GC laboratory. As a rapidly developing technique, the selection of the best approach for a particular application of SFE-GC is likely to require continued evaluation of new reports of SFE-GC methodology.

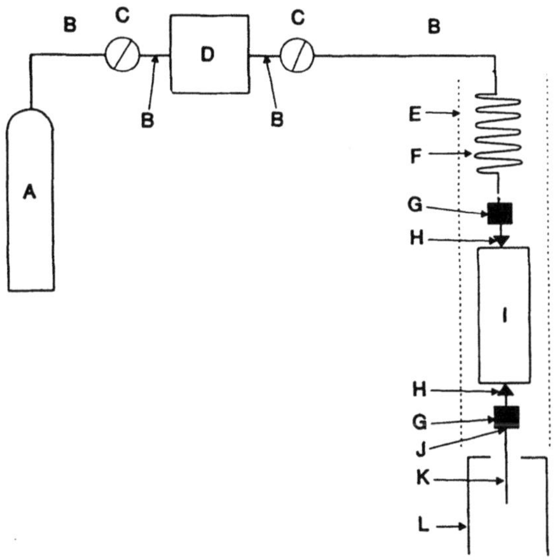

Figure 3.2 Schematic of a simple SFE-GC system showing all the required components. (Several manufacturers supply suitable components and specific suppliers are listed only for the reader's convenience.) Components are: (A) SFE grade extraction fluid source; (B) 1.5 mm o.d. stainless steel tubing (0.77 mm or smaller i.d.); (C) shut-off valves (SSI model 02–0120 or equivalent, Supelco, Bellefonte, PA); (D) SFE pump (see chapter 2); (E) SFE cell heater (see chapter 2); (F) approx. 0.5 m long coil of 1.5 mm stainless steel tubing for fluid preheater; (G) 1.5 mm × 1.5 mm tubing union (e.g. Parker or Swagelok brand); (H) finger-tight connectors (e.g. Slip-Free connectors from Keystone Scientific, Bellefonte, PA, USA); (I) SFE cell (see chapter 2); (J) restrictor connector ferrule (Supelco M2-A, Bellefonte, PA) which is used to replace the stainless steel ferrule in the outlet end of the tubing union 'G'; (K) 15–30 μm i.d. fused silica tubing restrictor (Polymicro Technologies, Phoenix, AZ); (L) GC injection port (see Figures 3.3 and 3.4).

3.3 Performing direct injection on-line SFE-GC

3.3.1 Basic instrumentation

As shown in Figure 3.2, the required instrumentation to perform direct injection SFE-GC includes a suitable pump, extraction cell and heater, a depressurization restrictor, suitable transfer lines and a conventional gas chromatograph. (The reader is referred to chapter 2 for a discussion of pump characteristics and alternatives.) Extraction cells are the same as those used for off-line SFE, although smaller cells (e.g. < 5 ml) have most often been used for on-line SFE-GC (for reasons discussed later). Depressurization restrictors for SFE-GC have most frequently been made from lengths of fused silica tubing (generally 15–40 μm i.d.) although suitable restrictors have also been made by crimping the end of a stainless steel tube to achieve the desired flow rate [6,11].

Just as two general styles of injection ports, split/splitless and on-column, are used for the majority of liquid solvent injections for capillary GC, direct

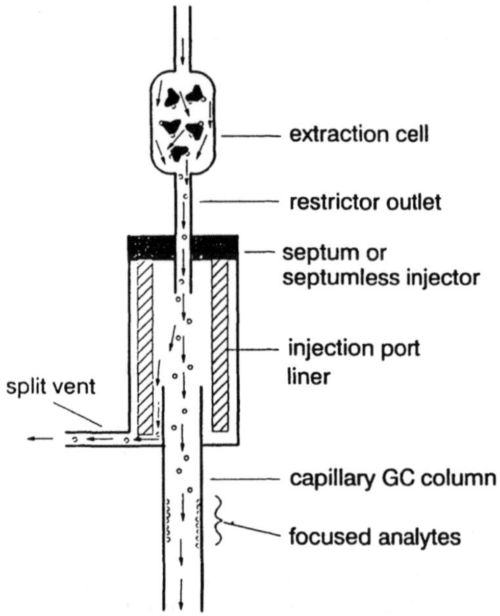

Figure 3.3 Schematic representation of a split SFE-GC interface.

injection SFE-GC methods have been developed using both split/splitless and on-column injection ports. Analytes in the SFE extracts can be transferred to either type of injection port using a heated transfer line [7] or by direct depressurization of the supercritical fluid into the injection port itself (Figures 3.3 and 3.4). When a heated transfer line is used, the depressurization occurs outside the gas chromatograph and the analytes are transferred in the gas phase by sweeping the heated transfer line with the GC carrier gas. However, the use of the heated transfer line (and the possibilities for analytes loss and contamination of the line) can be avoided by directly depressurizing the supercritical fluid in the GC injection port itself (Figures 3.3 and 3.4). With this approach, SFE-GC can be successfully conducted with little or no modification of conventional GC instrumentation, and (assuming a suitable pump is available) the entire apparatus can be assembled in a few hours or less.

3.3.2 Performing split SFE-GC

SFE-GC using a conventional split injection port has been by far the simplest and most reliable approach to SFE-GC that we have utilized, and this technique is strongly suggested as the first method to try. In addition to its inherent simplicity, one of the most important advantages of split SFE-GC is that the depressurization occurs in the heated injection port, which eliminates restrictor plugging from water and matrix components (as discussed in chapter 2) and also allows extractable matrix components that are not 'GC-able' to be deposited on the injection port liner instead of being introduced directly into the GC column as is the case for the on-column SFE-GC method discussed below. (Interestingly, we have encountered some particularly difficult samples that could not be extracted using off-line SFE because of restrictor plugging, that are quite easy to analyze using split SFE-GC because the depressurization of the supercritical fluid occurs in the heated injection port.) The analytes are then swept into the GC column and cryogenically trapped in the column's stationary phase until the extraction is completed. As with a conventional split injection of a liquid solvent, a proportion of the extracted analytes are also purged out of the split vent of the injection port (depending on the split ratio).

Converting a conventional capillary GC that is equipped with a split/splitless injection port for split SFE-GC can be accomplished in only a few minutes. All of the components needed to perform split SFE-GC are shown in Figures 3.2 and 3.3. The SFE-GC coupling simply involves mounting the SFE cell and heater directly above the injection port, and inserting the SFE outlet restrictor into the injection port liner so that the end of the restrictor is approximately 2 cm above the inlet of the GC column as shown in Figure 3.3. The restrictor must form a gas-tight seal with the injection port in order to ensure that the split ratio remains constant during the SFE step. This can be accomplished by carefully inserting the restrictor through the conventional GC septum (the end of the restrictor should be trimmed after it is inserted through the septum to ensure that the restrictor is not plugged with material from the

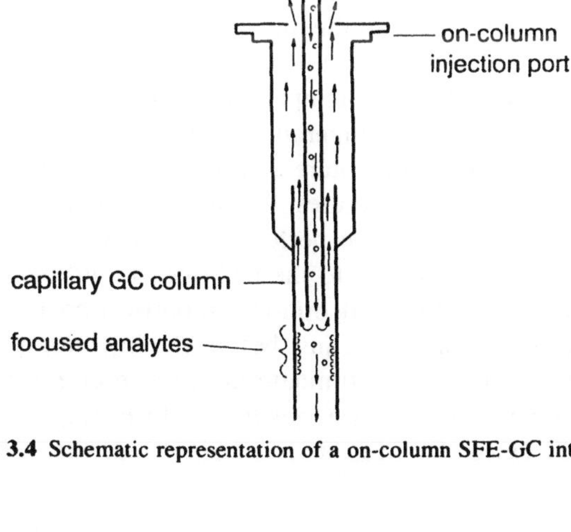

Figure 3.4 Schematic representation of a on-column SFE-GC interface.

septum). This method requires no modification of the GC injection port, but does require removal of the septum nut and septum each time a new restrictor is used. An attractive alternative that is routinely used in our laboratory is to attach a septum-less injector (Scientific Glass Engineering model SLI-M) to the top of the injection port [10]. This port, which is attached using a septum nut and graphitized ferrule, can be replaced with a conventional septum in a few seconds, thus allowing a single GC to be used for SFE-GC or conventional liquid solvent injections with only seconds required to convert between modes. The restrictor is passed through a cylindrical seal in the SGE septum-less injector into the injection port liner to provide a gas-tight connection between the injection port and the SFE restrictor. (We have performed the majority of our split SFE-GC studies using the split injector supplied by Hewlett-Packard with model 5880 or 5890 gas chromatographs, and the following discussions refer to the use of that injection port. However, other manufacturers' injection ports have also been used successfully. In addition, the components suggested in the following discussions are those that we have found to be reliable and inexpensive, although acceptable substitutes may be available from a number of suppliers. Specific suppliers are included only for the reader's convenience.)

The SFE pump is connected with 1.5 mm stainless steel tubing to a single shut off valve which is in turn connected to the extraction cell using finger-tight fittings. A length of the stainless steel tubing (approx. 50 cm) is coiled inside the heater prior to the extraction cell to prewarm the supercritical fluid to the extraction temperature. The extraction cell outlet restrictor is also connected to the extraction cell with finger-tight fittings and a capillary GC column ferrule. The cell and its heater are mounted in a clamp directly above the injection port so that the extraction cell restrictor can be inserted directly into the GC injection port without the use of transfer lines. A suitable cell heater can easily be made by evenly wrapping a length of metal tube (approx. 15–30 cm length to accommodate the extraction cell and preheating coil) with a heat tape followed by an external wrapping of insulation. The addition of thermocouple (silver soldered to the metal tube) and temperature control units also simplifies temperature control. (Since SFE extraction rates generally do not vary significantly with small temperature changes, the analyst need not be overly concerned with exact temperature control for the cell heater, and our experience indicates that temperature control plus or minus 5°C is adequate.) With the apparatus shown in Figures 3.2 and 3.3, no additional modification of the capillary GC needs to be performed with the possible exception of adding a cryogenic cooling unit to the GC oven to aid the trapping of more volatile analytes. One additional optional modification is to add a shut-off valve in the carrier gas inlet line just before the injection port. This valve is closed during the SFE step to eliminate the chance of extracted vapor phase analytes backing up into the carrier gas line and causing contamination of the carrier gas line. Once the SFE step is completed, the valve is opened to reintroduce the carrier gas flow to the injection port. Although we have not observed such contamination in SFE-GC systems without the shut-off valve, we have generally added the valve to eliminate any potential problems.

Once the apparatus shown in Figures 3.2 and 3.3 is assembled, the SFE-GC analysis is performed by the following steps:

(1) The extraction cell is filled, assembled, and attached to the restrictor and the inlet tubing from the pump.
(2) The GC oven is cooled to an appropriate trapping temperature.
(3) The extraction cell is placed in the tube heater, and the restrictor is inserted into the split injector so that the end of the restrictor is approximately 2 cm above the column inlet (Figure 3.3).
(4) The cell is pressurized and SFE is performed dynamically for a suitable time period (generally 10–30 min). During this time the extracted analytes are volatilized in the heated injection port and a fraction of them (depending on the split ratio) are swept into the GC column where they are cryogenically focussed in the column's stationary phase.
(5) The extraction flow is stopped and the SFE restrictor is withdrawn from the injection port.
(6) The gaseous extraction fluid is allowed to vent from the GC column for 1–3 min (approx. one void volume of the capillary column).
(7) The GC temperature program is begun and the GC analysis is completed in a normal manner.

3.3.3 Performing on-column SFE-GC

The first SFE-GC method to be reported was developed based on the observation that the fused silica restrictors used for off-line SFE closely resemble the fused silica needles commonly used for conventional on-column injections of liquid solvents [3,14]. With the use of restrictors that have sufficiently small outer diameters (e.g. 150 μm o.d.) on column SFE-GC can be performed by simply inserting the SFE restrictor into the capillary GC column through a conventional on-column injection port as shown in Figure 3.4. The length of the restrictor should be measured so that the end of the restrictor is well within the initial portion of the GC column that is cooled by the low oven temperature. (A 15-cm restrictor is appropriate for the style of on-column injector that is supplied with a Hewlett-Packard 5890 gas chromatograph.) With this approach, the supercritical fluid is depressurized to ambient pressure inside the GC column itself, and the analytes are cryogenically focussed in the GC column during the SFE step to provide narrow chromatographic bands. The steps for performing the on-column SFE-GC analysis are virtually identical to those listed above for the split technique, except that the restrictor is placed directly inside the capillary GC column (either 250 or 320 μm i.d.) rather than in the split injection port. With this arrangement, the analytes are deposited directly in the cooled stationary phase of the GC column while the majority of the gaseous extraction fluid is vented back through the on-column injection port as shown in Figure 3.4 (although a significant portion of the extraction fluid also travels through the GC column as discussed below). This is in direct contrast to the split technique, where a gas-tight seal with the restrictor is required to maintain a known split ratio during the extraction. With the on-column technique, however, a gas-tight seal between the restrictor and the on-column injection port cannot exist because the entire flow of the gaseous extraction fluid cannot be vented rapidly enough through the capillary GC column.

3.3.4 Choosing an SFE-GC method

While both on-column and split SFE-GC methods have been reported to yield good quantitative reproducibilities (e.g. 2–10% relative standard deviations for replicate SFE-GC analyses), and several examples of successful quantitative extraction and analyses have been reported using each method [1, 2, 4, 6, 10, 12–14, 16–18], the choice between split and on-column SFE-GC depends heavily on the sample characteristics and the need for maximum sensitivity. Comparisons between the methods are analogous to similar comparisons for conventional liquid solvent injections utilizing a split or on-column injector. On-column SFE-GC has an inherent sensitivity advantage, since every extracted analyte molecule is transferred to the GC column, while split SFE-GC vents a proportion of the analytes (depending on the split ratio). On-column SFE-GC also avoids exposing thermally labile analytes to the high injection port temperatures encountered using split SFE-GC. However, on-column SFE-GC often fails when the sample contains high concentrations of water because the water freezes into an ice plug at the outlet of the restrictor and blocks further extraction flow. In contrast, split SFE-GC yields somewhat lower sensitivity than on-column SFE-GC since only a fraction of the extracted analytes are introduced into the GC column (depending, of course, on the split ratio). However, split SFE-GC handles wet samples very well since the depressurization occurs in the hot (typically 300°C) injection port, which completely eliminates restrictor plugging from frozen water that occurs with on-column SFE-GC. While it is necessary to maintain the cryogenic trapping temperature above 0°C during split SFE-GC of wet samples to avoid plugging the capillary GC column with ice, quite wet samples (e.g. 20% water in sediment) can be analyzed using split SFE-GC [10].

While the major advantage of the on-column approach is that all analyte molecules are deposited in the GC column, it is unfortunately true that all extracted matrix molecules are also deposited in the GC column. In contrast, the split SFE-GC system allows non-volatile matrix components that are extracted to be deposited on the injection port liner, and is therefore more useful for the analysis of dirtier samples. As discussed below, split SFE-GC can also accommodate higher SFE flow rates (and therefore is useful for larger samples) than on-column SFE-GC. Considerations of these general characteristics (as well as the experience of analyzing several hundred samples by SFE-GC techniques) leads us to the conclusion that split SFE-GC is nearly always the method of choice, particularly for the first attempt at performing SFE-GC. However, when maximum sensitivity is needed from small samples (e.g. samples that are difficult to obtain in large quantities such as air particulate matter), on-column SFE-GC is clearly the method of choice. For example, if the detection limit of a GC detector is 1 pg for a particular analyte, on-column SFE-GC is capable of obtaining the impressive detection limits of ppb (ng/g) sensitivities from a 1-mg sample, or ppt (pg/g) sensitivities from a 1-g sample, while split SFE-GC sensitivities are poorer by a factor of about 10–100 (depending on the split ratio).

3.3.5 Optimizing SFE-GC methods

In addition to developing appropriate SFE conditions for a particular sample (discussed in chapter 2), a coupled SFE-GC method must balance the SFE conditions with the need to efficiently focus the analytes in the GC column during the extraction period. For example, the extraction of large (several grams) samples may require higher supercritical fluid flow rates, which makes the quantitative collection and efficient focusing of the extracted analytes in the GC system more difficult. With both direct injection SFE-GC methods, two major factors, SFE flow rate and cryogenic trapping temperature, are important for achieving efficient collection of the analytes in the GC column and, therefore, good chromatographic separations of the extracted analytes. Both on-column and split SFE-GC are limited by the flow rate of the depressurized SFE fluid that can be accommodated by the GC. With on-column SFE-GC, the maximum useful flow rate (as determined by the inner diameter of the restrictor) depends on the inner diameter of the GC column. For example, the effect of flow rate and GC column i.d. is shown in Figure 3.5 for the on-column SFE-GC analysis of a sediment contaminated with coal tar. When the SFE flow rate was maintained at approximately 0.5 ml/min (resulting in a gas flow of approx. 250 ml/min) using a 25 μm i.d. restrictor inserted into a 320 μm i.d. DB-5 column, the chromatographic peak shapes were quite good. However, when the flow rate was increased to 0.8 ml/min using a 30 μm i.d. restrictor, the peak shape degraded, and some double peaks could be observed (indicated by the arrows). Increasing the SFE flow rate completely destroyed the chromatographic peak shapes. For a 250 μm i.d. GC column, an SFE flow rate of approximately 0.5 ml/min yielded poorly formed and split peaks (Figure 3.5, bottom), and obtaining good peak shapes required lowering the supercritical N_2O flow rate to 150 μl/min using a 15 μm i.d. restrictor. These limitations imposed on the SFE flow rate by the inner diameter of the

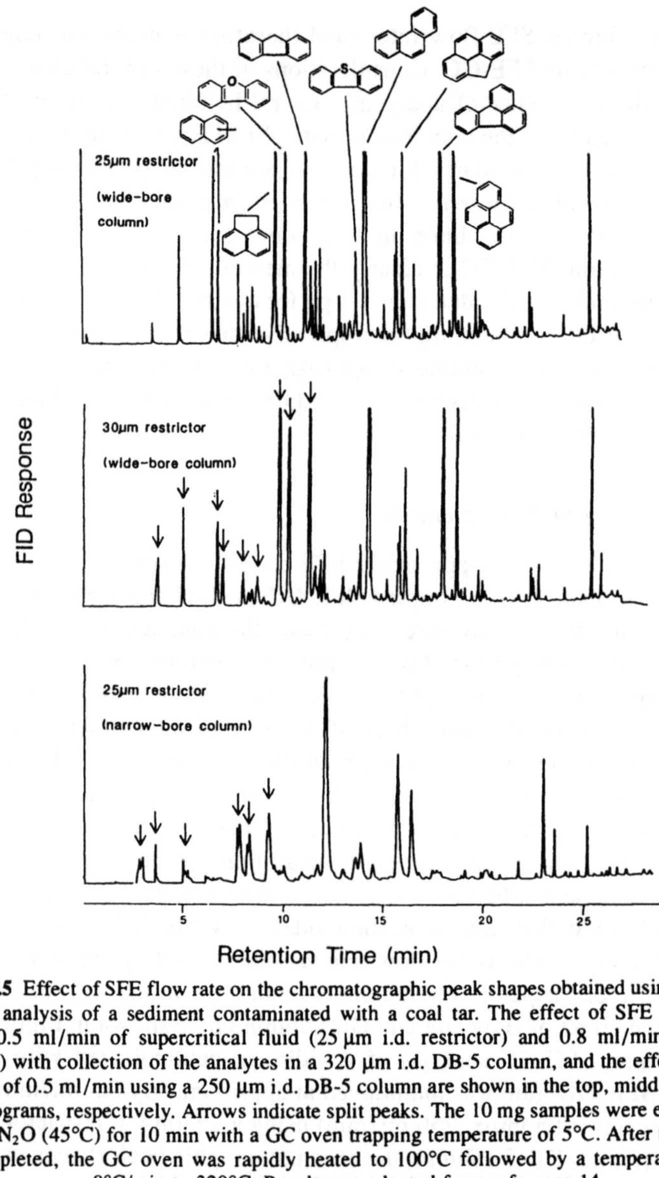

Figure 3.5 Effect of SFE flow rate on the chromatographic peak shapes obtained using on-column SFE-GC analysis of a sediment contaminated with a coal tar. The effect of SFE flow rates of approx. 0.5 ml/min of supercritical fluid (25 μm i.d. restrictor) and 0.8 ml/min (30 μm i.d. restrictor) with collection of the analytes in a 320 μm i.d. DB-5 column, and the effect of an SFE flow rate of 0.5 ml/min using a 250 μm i.d. DB-5 column are shown in the top, middle and bottom chromatograms, respectively. Arrows indicate split peaks. The 10 mg samples were extracted with 300 atm N_2O (45°C) for 10 min with a GC oven trapping temperature of 5°C. After the extraction was completed, the GC oven was rapidly heated to 100°C followed by a temperature ramp of 8°C/min to 320°C. Results are adapted from reference 14.

GC column demonstrate that on-column SFE is best suited to small (e.g. < 1 g) samples, simply because high extraction flow rates cannot be accommodated.

Split SFE-GC is less sensitive to the extraction flow rate, since the majority of the depressurized extraction fluid (and the extracted analytes) are vented through the split vent. The upper limit of SFE flow that can be used and still obtain good chromatographic peak shapes depends on the split ratio, that is, higher split ratios allow higher SFE flow rates. A reasonable starting point is to use a split ratio of about 1:20 to 1:100. For example, split SFE-GC analysis of a fuel contaminated sediment is shown in Figure 3.6 using 25, 30 and 40 μm i.d. restrictors (corresponding to approx. 0.6, 1.0 and 1.6 ml/min of supercritical CO_2, and 270, 450 and 750 ml/min of gaseous CO_2, respectively) and a 320 μm i.d. GC column with a split ratio of 1:50. Both 25 and 30 μm restrictors yielded chromatographic peak shapes that compare favorably with those obtained using a conventional split injection of a methylene chloride extract, however an SFE flow rate of 1.6 ml/min using the 40 μm restrictor resulted in poorer chromatographic peak shapes, particularly for the more volatile C12 to C20 alkanes. However, as discussed in chapter 2, flow rates

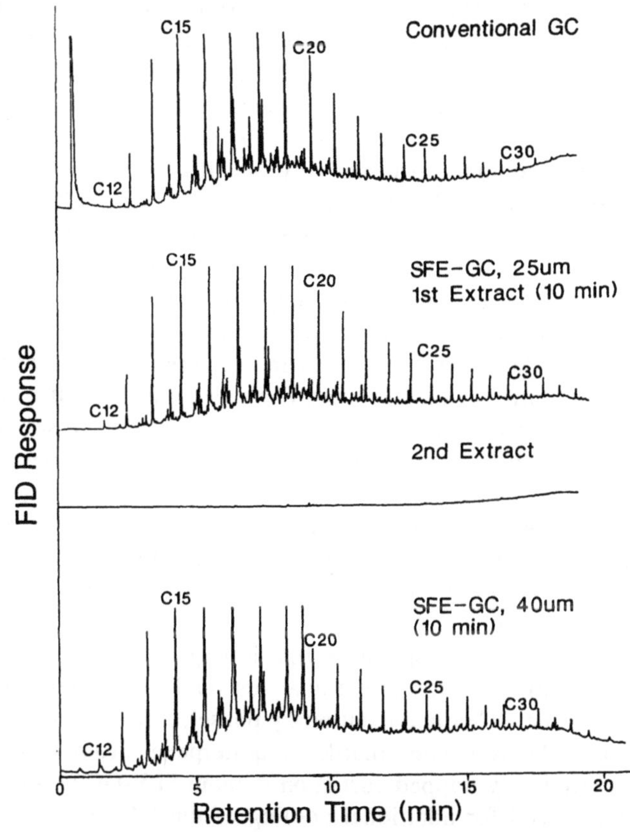

Figure 3.6 Comparison of the chromatograms generated using a conventional split injection of a methylene chloride extract (top) and split SFE-GC analysis (400 atm CO_2, 50°C, 10 min) of a fuel-contaminated sediment. The methylene chloride extract (2 μl) was injected in the split mode at an oven temperature of 120°C followed by a temperature ramp at 8°C/min to 320°C. SFE-GC was performed with an oven trapping temperature of 20°C using a 25 μm (middle) or a 40 μm (bottom) i.d. restrictor. After the SFE step was completed, the oven was rapidly heated to 120°C, followed by a temperature ramp at 8°C/min to 320°C. The numbers above the peaks indicate the length of the *n*-alkane chain. Results are adapted from reference 10.

of 1.0 ml/min are more than sufficient for many samples as long as the void space in the sample is swept in a reasonable time (every few minutes), and the relatively high flow rates that can be accommodated by split SFE-GC allows samples as large as 15 g to be analyzed [10].

It should be noted that the useful flow rate for either on-column or split SFE-GC can also be somewhat limited by some GC detectors. We have experienced no problems with ECD and FID detectors, however, the high gas flow rates that result in the GC column from either on-column or split SFE-GC can cause excessively high pressures in the source region of mass spectrometers, particularly with MS instruments with low pumping capacity (e.g. some 'bench-top' instruments). When SFE-GC/MS experiments are performed, it is helpful to reduce the gas flow into the source region during the SFE step. This can be accomplished by using lower SFE flow rates (if possible) for both on-column and split techniques. With on-column SFE-GC, reduced flows to the MS source can also be accomplished by using longer and narrower columns (e.g. choose a 60 m × 250 μm i.d. column rather than a 30 m × 320 μm i.d. column). Alternatively, the flow into the source can be reduced by installing a smaller diameter transfer line (e.g. 150 μm i.d. fused silica tubing) between the outlet of the GC column and the MS source [14]. This procedure does not reduce the sensitivity of the on-column SFE-GC method, since all of the analytes are still deposited in the GC column. The use of columns with smaller inner diameter (or transfer lines to the MS) also works with split SFE-GC, however, they result in an increased split ratio and thus poorer detection limits. In our laboratory, we generally use the GC column that is most appropriate for the target analytes and adjust the split ratio until the MS source pressure during the SFE step is acceptable. Because of the negative effects of high SFE flow rates on chromatographic peak shapes (discussed above) and the possible negative effects on flow-sensitive detectors (e.g. MS), the best general approach for both on-column and split SFE-GC is to use the lowest SFE flow rate that yields good extraction efficiencies in a reasonable period of time.

The cryogenic trapping temperature that is required for SFE-GC depends mostly on the volatility of the target analytes (and to a lesser extent on the characteristics of the GC column stationary phase). For example, as shown in Figure 3.7, the relatively non-volatile sesquiterpenes from rosemary spice (peaks C) are efficiently trapped during on-column SFE-GC at room temperature without the need for additional cooling of the GC oven. However, the volatile monoterpenes such as α-pinene (b.p. = 155°C, peak A) are poorly focused. At an oven temperature of – 10°C, bornyl acetate (b.p. = 223°C, peak B) is efficiently focussed, while an oven temperature of – 30°C is required for efficient trapping of the monoterpenes. It is interesting to note that the cryogenic trapping temperature can be too low using on-column SFE-GC, resulting in lower apparent extraction efficiencies as shown by the chromatogram in Figure 3.7 using a trapping temperature of – 50°C. These low re-

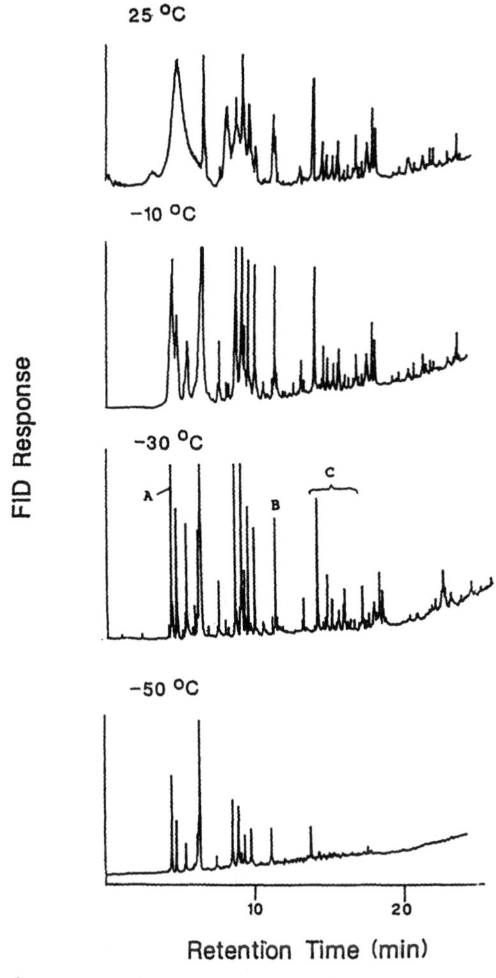

Figure 3.7 Effect of cryogenic (GC oven) trapping temperature on the chromatographic peak shapes obtained using on-column SFE-GC analysis of the spice rosemary. Each extraction was performed with 300 atm CO_2 at 45°C for 10 min with the analytes trapped in a 30 m × 320 μm i.d. (1 μm film thickness) DB-5 column. After the SFE was completed, the oven was heated rapidly to 70°C followed by a temperature ramp of 8°C/min to 320°C. Results are from reference 17.

coveries result because the cooling of the GC column from the expanding CO_2 combined with the low oven temperatures causes the CO_2 to freeze resulting in plugging of the restrictor outlet. The split SFE-GC technique does not suffer from this limitation since the supercritical fluid depressurizes in the heated injection port, and thus split SFE-GC can utilize lower trapping temperatures without restrictor plugging. (However, the analysis of samples with high amounts of extractable water using split SFE-GC limits the trapping temperature to >0°C to avoid freezing of the water in the GC column as discussed above.)

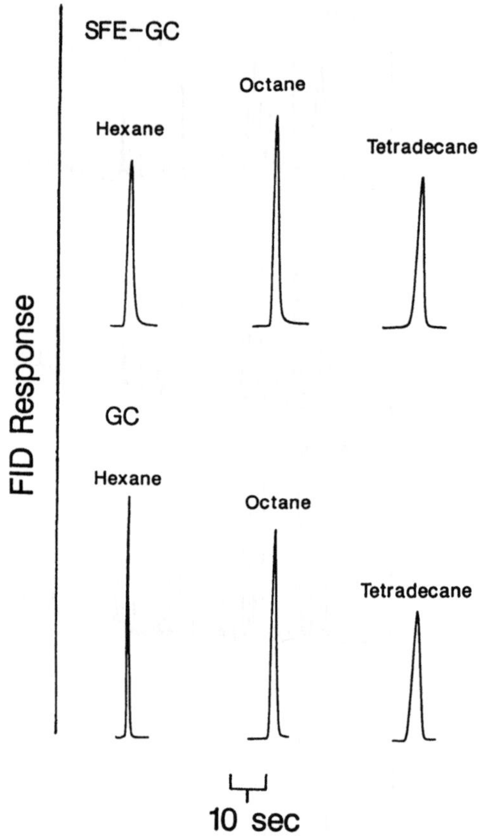

Figure 3.8 Comparison of the chromatographic peak shapes obtained for *n*-alkanes with split SFE-GC (top) and conventional autosampler split injection of a methylene chloride extract. SFE-GC was performed with 400 atm CO_2 at 50°C for 10 min with the GC oven held at – 50°C. After the extraction, the oven was heated rapidly to 35°C followed by a temperature ramp at 8°C to 300°C. Split injections of the *n*-alkanes in pentane solvent were performed at an oven temperature of 35°C followed by a temperature ramp at 8°C to 300°C. Results are from reference 10.

With proper selection of the extraction flow rate and the cryogenic trapping temperature, relatively volatile analytes can be efficiently trapped using either on-column or split SFE-GC. As shown in Figure 3.8, split SFE-GC yields reasonable peak shapes for species as volatile as hexane using a standard GC column (J&W DB-5, 30 m × 320 μm i.d., 1 μm film thickness), when compared to the peak shapes obtained on the same column using a split autosampler injection. On-column SFE-GC yields similar trapping efficiencies with species at least as volatile as *n*-octane [18]. In addition, both SFE techniques yield chromatograms that compare favorably with those generated using conventional liquid solvent injections of methylene chloride extracts as shown in Figure 3.6 by the split analysis of a fuel-contaminated sediment, and in Figure 3.9 by the on-column analysis of the spice basil.

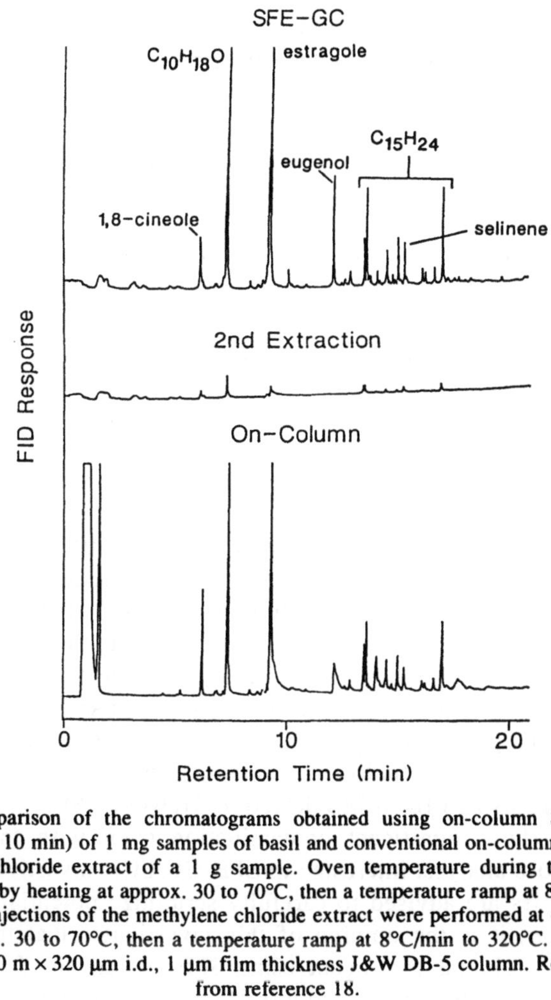

Figure 3.9 Comparison of the chromatograms obtained using on-column SFE-GC analysis (300 atm CO_2 for 10 min) of 1 mg samples of basil and conventional on-column injections (1 µl) of a methylene chloride extract of a 1 g sample. Oven temperature during the SFE step was − 30°C, followed by heating at approx. 30 to 70°C, then a temperature ramp at 8°C/min to 320°C. The on-column injections of the methylene chloride extract were performed at 40°C followed by heating at approx. 30 to 70°C, then a temperature ramp at 8°C/min to 320°C. Separations were performed on a 30 m × 320 µm i.d., 1 µm film thickness J&W DB-5 column. Results are adapted from reference 18.

3.3.6 Selection of extraction fluids for SFE-GC

The choice of fluids for SFE-GC is limited by the need for volatility (as the depressurized fluid) in addition to the general characteristics of SFE fluids discussed in chapter 2. Not surprisingly, the majority of SFE-GC work has been conducted using pure CO_2, although other fluids that are gases at ambient conditions including N_2O, SF_6 and low molecular weight alkanes could be useful. In particular, N_2O has been shown to yield better recoveries of some analytes (e.g. higher molecular weight PAHs from sediment [10] when used for SFE-GC N_2O is an oxidant, however, and SFE of samples with high organic content at high temperatures could represent an explosion hazard and should be avoided.) Supercritical CO_2 is certainly a good first fluid to try during the development of an SFE-GC method, and has the advantage that (as a very rough generalization) analytes extracted with pure CO_2 under normal SFE conditions (e.g. 300–500 atm) tend to be amenable to GC analysis, and organics that are not volatile enough to be analyzed by GC tend not to be extracted with pure CO_2. Unfortunately, triglyceride fats are an important exception to this generalized rule in that they extract readily in CO_2, but are not easily analyzed using capillary GC.

As discussed in chapter 2, it is often necessary to add organic polarity modifiers to CO_2 to obtain quantitative recovery of many analytes, a procedure that makes on-line SFE-GC techniques difficult to perform since modifiers (e.g. methanol) are generally not gases at ambient conditions. However, recent reports [11,19] have demonstrated the use of small amounts of modifiers including methanol, ethanol, hexane and CS_2 with split SFE-GC. Although the resultant chromatograms show a large 'solvent' peak for the organic modifier, the extracted analytes displayed reasonable chromatographic peak shape as shown in Figure 3.10.

The potential of SFE-GC to yield very good sensitivities by transferring every extracted analyte molecule to the GC injector also puts stringent requirements on the purity of the extraction fluid, since every contaminant molecule present in the extraction fluid will also be transferred to the GC. It is extremely important to utilize the highest purity fluids available to avoid any contamination. At least two suppliers (Scott Specialty Gases and Air Products) offer 'SFC' or the even purer 'SFE' grade CO_2. In our experience, these grades are sufficiently clean that virtually no contamination can be detected by on-column SFE-GC using FID or MS detection. However, SFE-GC with an ECD detector may show significant contamination even with these ultra-pure grades, and those analysts interested in SFE-GC/ECD investigations at maximum sensitivities are advised to carefully scrutinize potential SFE gas suppliers. However, it is the opinion of this investigator that the majority of contamination problems that occur using SFE-GC are the fault of the SFE pump and associated plumbing, and less often the fault of the fluid purity. We have found that using very simple systems with a minimum of components is the best approach to avoiding contamination problems. Metal to metal connections should be used exclusively, and sealants such as Teflon tape are especially to be avoided. If possible, it is also useful to designate one pump as a 'clean' pump, and never use that pump for any other fluid except ultra-pure CO_2 to eliminate the chance of contaminating pump seals.

Whenever the SFE-GC chromatograms show unacceptable contamination, the source must be determined. A simple test to determine whether the fluid or the SFE system is the contamination source is to completely eliminate the pumping system and all valves, and to simply vent the source CO_2 directly into the GC system using the tank pressure (i.e. connect the CO_2 cylinder directly to an appropriate restrictor and insert the restrictor into the GC injection port as if an SFE-GC experiment was being conducted).

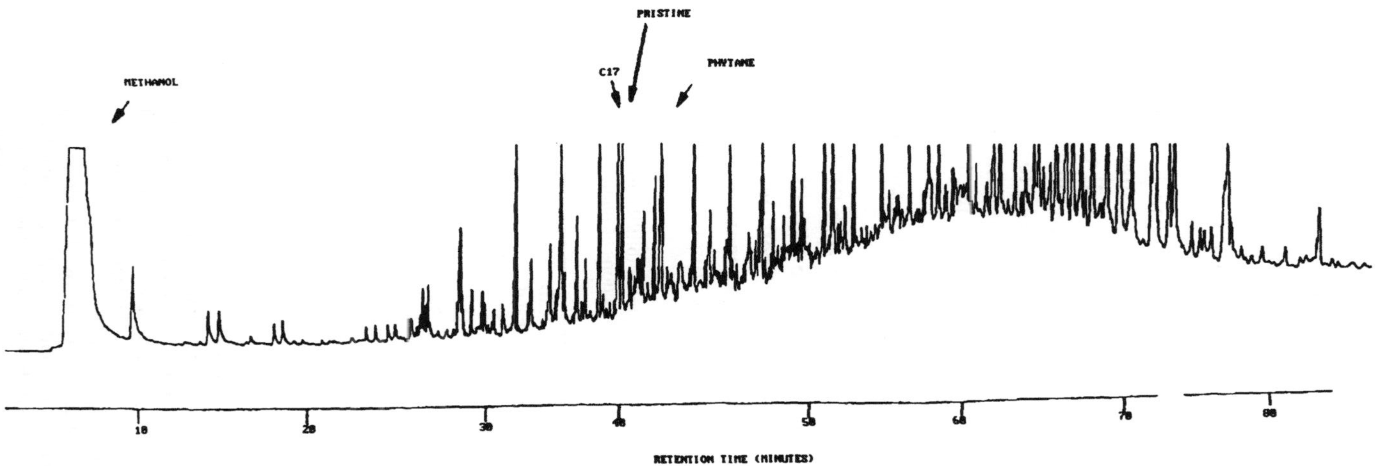

Figure 3.10 Split SFE-GC analysis of shale rock using on-line SFE-GC with 50 μl methanol modifier added to the sample cell. SFE was performed for 30 min static followed by 10 min dynamic with 350 atm CO_2 at 60°C. Results are adapted with permission from reference 19.

3.3.7 *Quantitative SFE-GC*

Calibration for both on-column and split SFE-GC has been accomplished using two methods. First, a few microliters of a standard solution can be added to an inert matrix (e.g. sand, glass beads or the extraction cell frit) in the extraction cell, and the standard then can be analysed using SFE-GC in a manner identical to that used for samples. (Care must be taken to avoid the loss of any volatile analytes.) Second, a standard can be injected using a conventional syringe into the same injection port as that used for the SFE-GC analyses (i.e. an on-column injection for on-column SFE-GC and a split injection for split SFE-GC). While both methods have been shown to give equivalent results, we routinely perform both calibrations to ensure the quality of the quantitative data. Additionally, an appropriate internal standard should be added to each calibration standard, and should also be spiked into the extraction cell with the sample prior to SFE-GC analysis (the analyst must be certain, however, that the SFE conditions will yield quantitative recovery of the spiked internal standard).

Several examples of using both on-column and split SFE-GC for quantitative analyses have been reported in the literature [4,6,10,12–14,16–18], and once the SFE conditions have been developed, both techniques have been shown to yield good quantitative comparisons to conventional extraction and analysis techniques. The reproducibility of on-column and split SFE-GC tech-

Table 3.1 Comparison of peak area reproducibilities for direct injection SFE-GC versus conventional injections of liquid solvent extracts

Analyte	% RSD SFE-GC	% RSD Conventional
On-column SFE-GC of basil[a]		
1,8-Cineole	17	11
Estragole	9	11
β-Selinene	9	15
Sesquiterpene isomer	8	10
Split SFE-GC of fuel-contaminated river sediment[b]		
Dodecane	2	6
Hexadecane	12	11
Docosane	3	5
Octacosane	5	5

[a] % RSD values were based on quadruplicate on-column SFE-GC analyses of 1-mg samples of basil, or on quadruplicate 1-μl on-column injections of a single methylene chloride extract of basil. SFE-GC conditions are given in the legend of Figure 3.9. Results are adapted from reference 18.

[b] RSD were based on triplicate split SFE-GC analyses of 1.3-g samples of the sediment, or on triplicate 1-μl split injections (using an autosampler) of a single methylene chloride extract. SFE-GC conditions are given in the legend of Figure 3.6. Results are adapted from reference 10.

niques have also been reported to be acceptable (e.g. typically 2–10% RSDs depending on the sample) as long as the samples to be extracted are representative of the bulk sample (this is of particular concern for the small samples used for on-column SFE-GC). As shown in Table 3.1 for the on-column SFE-GC analysis of the spice basil (Figure 3.9), and the split SFE-GC analysis of a fuel-contaminated sediment (Figure 3.6), replicate SFE-GC analyses yielded reproducibilities that were similar to those obtained by multiple GC analyses of a single methylene chloride extract. Note that SFE-GC analyses include any errors in SFE extraction and the cryogenic trapping step, and any inhomogeneity of the small samples used (1.3 g for the split SFE-GC analysis of the contaminated sediment and 1 mg for the on-column SFE-GC analysis of the basil), in addition to any irreproducibility in the GC analysis. In contrast, the conventional liquid solvent injections only include errors associated with the GC analysis since a single extract of each sample was injected multiple times. As shown in Table 3.1, both on-column and split SFE-GC give reproducibilities similar to those obtained by the liquid solvent injections.

3.4 Summary

SFE-GC is an attractive approach to coupling the extraction, concentration and chromatographic steps for the analysis of samples containing analytes that can be analyzed using capillary GC. Quantitative analyses can be performed and only minimal modification of conventional GC instrumentation is needed. However, it is important for the analyst to evaluate the reasons for selecting on-line SFE-GC over off-line SFE techniques. Off-line is inherently simpler for the novice to perform, since only the SFE (and analyte collection) step needs to be understood, and subsequent analyses can be performed using conventional liquid solvent injection techniques. SFE-GC appears to have advantages when relatively volatile analytes (e.g. hexane) need to be determined, particularly when such species are difficult to collect using off-line methods. However, present SFE-GC techniques are not useful for samples that contain high concentrations of extractable, but not 'GC-able' matrix components such as fats, while off-line extracts can be fractionated using normal procedures for liquid solvent extracts. SFE-GC does have the advantage that an entire analysis including extraction, sample concentration and GC separation can normally be completed in 30–60 min. However, the daily sample throughput can be higher using off-line SFE, since SFE-GC requires that the GC be used for a sample collection device (rather than performing chromatographic separations) during the SFE extraction, whereas several off-line extracts can be loaded into an autosampler for unattended GC analysis. The greatest potential advantage of SFE-GC is that, since all of the extracted analytes are transferred to the GC system, much higher method sensitivities can be obtained. With off-line SFE, sensitivities are limited by the fact that only about 1 μl of the collection solvent is generally injected into the GC. It should be emphasized that present SFE-GC techniques are certain to develop and become more reliable, simpler, and more amenable to automated analysis. The present SFE-GC methods discussed in this chapter clearly demonstrate the potential for on-line SFE-GC to provide rapid, sensitive, reliable and quantitative analyses of a variety of samples.

Acknowledgements

The authors would like to thank the US Environmental Protection Agency, the New Jersey Department of Environmental Protection (Office of Science and Research), Shell Oil Company (USA), British Petroleum (USA) and the US Department of Energy for support of their SFE investigations. Instrumental support from ISCO (USA) is also gratefully acknowledged.

References

1. R. W. Vannoort, J.-P. Chervet, H. Lingeman, G. J. DeJong and U. A. Th. Brinkman, *J. Chromatogr.* **505** (1990) 45.
2. S. B. Hawthorne, *Anal. Chem.* **62** (1990) 633A.
3. S. B. Hawthorne and D.J. Miller, *J. Chromatogr. Sci.* **24** (1986) 258.
4. F. I. Onuska and K.A. Terry, *J. High Resolut. Chromatogr.* **12** (1989) 527.
5. B. W. Wright, S. R. Frye, D. G. McMinn and R. D. Smith, *Anal. Chem.* **59** (1987) 640.
6. J. M. Levy and J. P. Guzowski, *Fresenius Z. Anal. Chem.* **330** (1988) 207.
7. J. M. Levy, J. P. Guzowski and W. E. Huhak, *J. High Resolut. Chromatogr. Chromatogr. Commun.* **10** (1987) 337.
8. J. H. Raymer and E. D. Pellizzari, *Int. J. Environ. Anal. Chem.* **43** (1991) 151.
9. M. W. F. Nielen, J. T. Sanderson, R. W. Frei and U. A. Th. Brinkman, *J. Chromatogr.* **474** (1989) 388.
10. S. B. Hawthorne, D. J. Miller and J. J. Langenfeld, *J. Chromatogr. Sci.* **28** (1990) 2.
11. R. J. Houben, H.-G. M. Janssen, P. A. Leclercq, J. A. Rijks and C. A. Cramers, *J. High Resolut. Chromatogr.* **13** (1990) 669.
12. C. K. Huston and H. J., *J. Agric. Food Chem.* **39** (1991) 1229.
13. T. J. Nielsen, I. M. Jägerstad, R. E. Öste and B. T. G. Sivik, *J. Agric. Food Chem.* **39** (1991) 1234.
14. S. B. Hawthorne and D. J. Miller, *J. Chromatogr.* **403** (1987) 63.
15. M. Lohleit and K. Bächmann, *J. Chromatogr.* **505** (1990) 227.
16. J. M. Levy and A. C. Rosselli, *Chromatographia* **28** (1989) 613.
17. S. B. Hawthorne, M. S. Krieger and D. J. Miller, *Anal. Chem.* **60** (1988) 472.
18. S. B. Hawthorne, D. J. Miller and M. S. Krieger, *J. Chromatogr. Sci.* **27** (1989) 347.
19. J. M. Levy, E. Storozynsky and M. Ashraf-Khorassani, *Recent Advances in Supercritical Fluid Technology* ed. M. E. McNally, American Chemical Society Symposium Series, Washington, DC, (1991).

4 Coupled supercritical fluid extraction-capillary supercritical fluid chromatography

N.J. COTTON, M.W. RAYNOR and K.D. BARTLE

4.1 Introduction

Analytical chemists are always seeking simple more efficient and less time-consuming sample preparation and analysis techniques. As a result there has been a steadily growing interest in the use of supercritical fluids for extraction and chromatography.

A good solvent for extraction should be selective so that it dissolves the desired analytes to a greater degree than other constituents in the sample matrix. It should be unreactive and stable, preferably non-toxic and from an economic point of view non-corrosive to equipment and inexpensive to buy. Many of these requirements are met by supercritical fluids such as carbon dioxide.

The unique properties of supercritical fluids that are of interest for extraction are the mass transfer properties, density and viscosity. These parameters have been discussed in chapters 1 and 2 and give rise to several major advantages over liquid extraction for sample preparation prior to chromatographic analysis. These advantages include:

(1) Improved extraction efficiencies in many cases with respect to Soxhlet extraction, resulting in reduced extraction times. Supercritical fluids have more rapid mass-transfer properties than liquids, and therefore penetrate and extract components in sample matrices more quickly and efficiently. In contrast to Soxhlet extraction, which requires several hours or days to perform, SFE can yield quantitative extraction in as little as 10 min [1]. On some compound groups, SFE is reported to give better recoveries than liquid–liquid extraction [2].

(2) The possibility of adjusting the solvent strength of the fluid. As the solvating power of a supercritical fluid is directly related to its density, it can be tuned to the analysis as required. For example, the sample can be extracted exhaustively at constant high density to provide quantitative analysis, or at various pressures or with different mobile phases to fractionate the sample.

(3) The extraction of heat sensitive or reactive compounds at low temperatures.

(4) The use of extraction solvents such as CO_2 which are non-toxic and inexpensive.
(5) The removal of the supercritical fluid from the extracted analytes by decompression.
(6) The possibility of on-line chromatographic analysis after extraction. As the extraction solvent is easily removed while the analytes of interest are trapped for subsequent separation, trace analysis is possible with on-line GC or SFC. In direct coupling, the solvent peak is eliminated and the analysis of compounds that elute with the solvent becomes possible. Further, on-line coupling reduces sample handling, subsequent possibility of sample loss and contamination of the sample.

Obviously many of the advantages discussed above are also applicable to SFC. SFC is a particularly good technique for analysing thermally labile, reactive or involatile materials, because separations are performed at low temperatures. The higher diffusion coefficients of solutes in supercritical mobile phases in contrast to those in liquids allow the use of more efficient open tubular capillary columns. Further, supercritical fluids such as CO_2 are

Table 4.1 Comparison of reported areas of on-line SFE-SCFC and SFE-GC

Sample matrix	Analysis	SFE-SFC	SFE-GC	Reference
Industrial samples				
Smokeless powders	Explosives, propellants	Yes	No	4
Commercial plastics	Polymer additives, oligomers	Yes	No	5,6,7,8,9
Biological samples				
Animal and plant tissue	Steroids, vitamins, drug residues	Yes	Yes	10,11,12,13,14
Water, urine	Phenols, phosphonates	Yes	No	15,16
Foods/natural products				
Butter, cheese, oils	Fatty acids, triglycerides	Yes	No	17
Coffee	Flavour compounds, caffeine	Yes	No	17,18,19,3
Spices	Flavour and fragrance compounds	Yes	Yes	20,21,18
Fruit peels, leaves, wood	Terpenes, aldehydes, alcohols	Yes	Yes	20,21,22
Environmental samples				
Coal, coal tar, treated wood	PAHs, sesquiterpenes,	Yes	Yes	18,23
Soil and sedimentary samples	PCBs, PAHs pesticides	Yes	Yes	24,25,22
Air and exhaust particulates	PAHs, hydrocarbons, nicotine	No	Yes	26,20,27,18
Sorbents				
Tenax-GC, charcoal	PCBs, PAHs, hydrocarbons	Yes	Yes	28,29,30
Silica, alumina, LC material	PAHs, hydrocarbons	Yes	Yes	26,31,23,32
Polyurethane foam	Air, water pollutants, PAHs, phenols, hydrocarbons	No	Yes	20,33

Table 4.2 Selected applications of SFE-CSFC

Sample and extracts	SFE conditions	Trapping/sampling conditions	CSFC conditions	Detector	Quantitative	Reference
PAC and other aromatic compounds extracted from vacuum still bottoms	Pentane 210°C, 20–200 atm; static SFE	Aliquot sampling	Pentane 210°C, 30–45 atm at 0.2 atm/min, 20 m × 50 μm i.d. SB-phenyl-50	Fluorescence	No	39
Extractable compounds in cheese, butter coffee, tobacco and camomile	CO_2 40°C, 100 bar; static SFE	Aliquot sampling	CO_2 isobaric, 130 bar, temp program to 220°C, 14 m × 250 μm i.d. supelcowax 10	FID	No	17
PAC extracted from a coal tar pitch	CO_2 35–50°C, 70–200 atm 30 min; dynamic SFE	Trapped in rotor groove of cooled injection valve (200 nl)	CO_2 110°C, 0.25–0.74 g/ml at 0.006 g/ml per min, 10 m × 50 μm i.d. SB-biphenyl-30	FID	No	44
Caffeine from coffee, oligomeric compounds from polyethylene, hydrocarbons from spiked soil	CO_2 25–60°C, 50–250 bar; dynamic SFE	Extracted compounds deposited in 0.1 μl rotor groove of injection valve	CO_2 150°C, 107–423 bar at 7 bar/min, 10 m × 50 μm i.d. biphenyl-methyl-silicone phase	FID	No	19
Caffeine from decaf-feinated coffee sample	CO_2 50°C, 14 MPa, 10 min; dynamic extraction	Aliquot sampling	CO_2 80°C, 0.25–0.66 g/ml, 10 m × 50 μm i.d. DB-5 (0.2 μm film)	FID	No	14
Ouabain in biological assay	CO_2 80°C 400 atm, 30 min; dynamic extraction	Cryogenic trapping in 150 μm capillary; fraction collection	CO_2 80°C, 200–400 atm, 3 m × 50 μm i.d. SB-methyl	FID	No	11
Terpenes, limonene, carvone etc. from grapefruit oil	CO_2 70°C; dynamic extraction at 0.356 g/ml, 12 min	− 10 to − 65°C cryogenic trapping in deactivated capillary	CO_2 150°C 0.1–0.61 g/ml per min, 10 m × 100 μm i.d. SB-phenyl-5	FID	No	22

Table 4.2 *(continued)*

Sample and extracts	SFE conditions	Trapping/sampling conditions	CSFC conditions	Detector	Quantitative	Reference
Pesticides (diuron, linuron) in soil and wheat grain	CO_2 and modified CO_2 with MeOH and EtOH in cell at 120°C, 0.3–0.7 g/ml density; static and dynamic SFE	Off-line collection vessel or 200 nl aliquot injected into CSFC	CO_2 120°C, 0.2–0.7 g/ml 10 m × 100 μm i.d. SB-Cyanopropyl-25 or 10 cm × 1 mm C18 packed column	Liquid scintillation counting FID	Yes	24
C_{12}–C_{20} hydrocarbons and polymer additives from polypropylene	CO_2 90°C, 300 atm, 30 min; dynamic extraction	– 40°C cryogenic trap with: (1) deactivated fused silica; (2) OV-1 coated fused silica; (3) 250 μm fused silica packed with ODS	CO_2 90°C, 80–400 atm at 5 atm/min, 10 m × 100 μm i.d. octyl (0.5 μm dp)	FID	No	5
Cuticular hydrocarbons and waxes from fruit beetle, triglycerides and fatty acids from cottonseed kernel	CO_2 45°C, 200 atm, 1 min; dynamic extraction	Cryofocusing using expanding CO_2	CO_2 100°C, 50–290 atm, 10 m × 50 μm i.d. SB-methyl or SB-biphenyl-30	FID	No	47
Additives in polyethylene	CO_2 50°C 100–atm at 150 atm/min, 5 min; dynamic extraction	– 30°C cryrofocusing onto head of column	CO_2 140°C 100–200 atm at 3 atm/min then to 400 atm, 10 m × 100 μm i.d. SB-biphenyl-30	FID, FTIR	No	8
Essential oils, inerol, geranyl acetate from dragonhead herb	CO_2 60–100°C, 140 atm and CO_2 80°C 0.2–0.65 g/ml; static extraction	Not reported	CO_2 180°C 0.2–0.67 g/ml at 0.014 g/ml per min 10 m × 50 μm i.d. SB-octyl-50	FID	No	48

Waxes, fatty acids amides and other coating materials from polyester sewing thread	CO_2 70°C 35 MPa, 7 min; dynamic extraction	Cryogenic trap cooled with expanding CO_2	CO_2 130°C 11–41 MPa at 0.6 MPa/min, 10 m × 50 µm i.d. SB-biphenyl-30	FID	No	49
Polymer additives in polypropylene pellets, nylon pellets and poly (etherether ketone) granules	CO_2 70°C 50–400 atm; dynamic extraction from 2–10 min	Cryogenic trap cooled with expanding CO_2	CO_2 120°C 100–400 atm at 5 atm/min, 10 m × 50 µm i.d. SB-methyl	FID	Yes	6
Pesticide traces in rendered fats and soya oil	CO_2 100°C 0.2–0.6 g/ml over 15 min; dynamic extraction	SF solid phase extraction cryofocusing after SFC using 2 mm × 35 mm i.d. C_{18} ODS packed column	CO_2 100°C 0.2–0.6 g/ml 20 m × 50 µm i.d. SB-phenyl-50	FID	Yes	50
Explosives and propellants in smokeless powders and soil (spiked samples)	CO_2 75°C 350 atm, 150 ml/min flow of CO_2; dynamic extraction for 5–60 min	20 mm × 2 mm i.d. LC pre-column trap packed with amino silica	CO_2 75°C 60–350 atm at 8 atm/min, 10 m × 50 µm i.d. SB-biphenyl-30	FID UV ECD	No	4
Oligomers in poly (ethylene-terephthalate) films	CO_2 70°C 400 atm; 30 min dynamic extraction	Cryogenic trapping	CO_2 130°C 100–400 atm at 10 atm/min, 10 m × 50 µm i.d. SB-biphenyl-30	FID	Yes	7

also compatible with flame-based detectors and with Fourier transform infrared and mass spectrometry.

SFE and SFC as individual techniques have been used for many years. However, it was not until 1985 that an on-line SFE-SFC system was reported by Sugiyama [3]. Tables 4.1 and 4.2 list the SFE-SFC studies and applications that have been reported in the literature since then. These studies, in keeping with the title of this chapter, are restricted to those that report on-line extraction and capillary SFC analysis.

4.2 Samples for which SFE-CSFC is applicable

Table 4.1 shows a comparison of the application areas of SFE-SFC and SFE-GC. It is clear that if the analytes are thermally stable and volatile that GC is the best separation technique to use. Many flavour and fragrance compounds in complex food samples should therefore be analysed by SFE-GC. The same is true of PCBs, pesticides and PAHs in environmental samples.

When the sample contains thermally labile or reactive compounds, SFE-SFC is recommended. The procedure is excellent for thermally unstable polymer additives in commercial plastics or for fatty acids and triglycerides in foods, etc. which cannot be analysed by GC very easily without derivatization. Natural products such as those containing terpene compounds or hops which contain highly reactive bitter compounds such as humulone and lupulone must also be analysed by SFC or HPLC as rearrangement can easily occur at elevated temperatures. Other application areas specific to SFC include the analysis of explosives and certain steroids, vitamins and other drug residues in biological samples. SFE-CSFC finds important applications in environmental science, both for the analysis of pollutants in matrices such as soil and sediments, and for the extraction of sorbents on which pollutants in air and water have been selectively adsorbed.

4.3 Influence of the sample matrix

The factors affecting the extractability of analytes during SFE have been fully discussed in chapter 1, where the important influence of sample matrix effects have been discussed: particle size and shape, adsorption effects, and the presence of water can all have a vital bearing. While a reduction in particle size leads to more rapid extraction, as predicted by the hot-ball model [34], care must be taken to avoid large pressure drops across the extraction cell and blockages arising from compaction of the material being extracted.

Matrix–solute interactions can severely reduce extraction efficiency and although well documented are, at present, incompletely understood. Thus Hawthorne [35] has shown how dioxins and polycyclic aromatic compounds may be so strongly adsorbed on matrices such as fly ash that complete recovery is impossible with conventional supercritical solvents. Recoveries from the adsorbent trapping materials are also highly variable. Wright *et al.* [36] compared the use of XAD-2 resin, polyurethane foam (PUF) and activated carbon as adsorbents for PAH; high recoveries were obtained for low molecular compounds with carbon dioxide from XAD-2 and PUF, but these were progressively reduced with increasing molecular weight. Tenax-GC was found to be a suitable material for the trapping and subsequent recovery by SFE of volatile compounds. Recoveries from carbon were negligible unless the CO_2 was modified. The choice of suitable adsorbent materials represents a compromise between break-through volume and the conditions required for subsequent extraction [37].

The presence of water in the matrix presents considerable problems during SFE-SFC. While the direct extraction of aqueous media has been described in SFE with coupling to packed column SFC [15, 16], it proved necessary to separate water from the supercritical fluid containing phenol analytes by means of a phase separator. The smaller capacities of capillary columns has made SFE-CSFC on aqueous samples difficult if not impossible.

The presence in the matrix of interfering compounds that are also soluble in the supercritical fluid, may also pose problems. However, by careful control of extraction pressure so as to limit co-extraction, it is possible to analyse for low levels of compounds of interest. King [37] has made use of the known solubilities of triglycerides to extract pesticides at the 100 ppm level from lard, while Nam *et al.* [38] demonstrated extraction of ppb level pesticides from a fish–lipid matrix.

4.4 Instrumentation

The instrumentation required for on-line SFE-CSFC can range from relatively simple systems to complicated arrangements involving ovens, switching valves and multiple pumps. However, in most systems there is an extraction cell which is temperature controlled or held in a thermostated oven, a switching valve, a trap which is usually cryogenically cooled, a chromatographic oven housing the capillary column, and a detector. The fluid may be delivered to both the extraction cell and capillary column using a single syringe pump. In some cases, when different conditions are required for extraction and chromatography simultaneously, a dual pumping system is required.

4.4.1 Static extraction

For on-line SFE-CSFC under static conditions, the sample to be extracted is held in a cell under pressure to equilibrate for a certain period of time before an aliquot of the extract is introduced into the capillary SFC column. A closed-

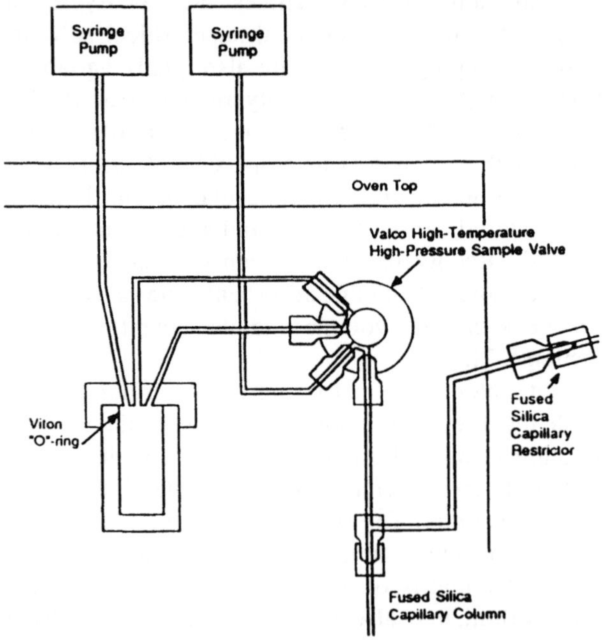

Figure 4.1 First reported static SFE-CSFC of Jackson *et al.*, closed-loop system (reprinted with permission from reference 39).

or an open-loop system may be used. Figure 4.1 shows the on-line (closed-loop) static SFE-CSFC investigated by Jackson *et al.* [39]. Two pumps were used, one for extracting the sample, and the other for performing the separation. The sample was sealed in a 1 ml stainless steel vial and pressurized with pentane at 210°C and 60 atm. After 30 min, an aliquot was injected into the SFC column and the analysis started.

The major advantage of this procedure is that small aliquots of the extract can be taken for consecutive analysis with virtually no difference in the extraction profile. However, a major disadvantage is that the solute containing extraction fluid has to reach equilibrium and diffuse out of the cell and into the injection valve before sampling is made. Wheeler and McNally have investigated static SFE-SFC in detail and found that it can take many hours before complete equilibrium is attained [24]. Thus, some sort of recirculating system could be used to reach equilibrium in a shorter time. Unfortunately recirculating pumps are not suitable for this purpose as they can easily become contaminated.

A solution is to incorporate a low flow restrictor connected to a switching valve in-line after the injector, as shown in Figure 4.2, so that extraction fluid can be drawn through the loop when required [24]. This is known as open-loop SFE, and with this configuration one also has the opportunity of passing the sample through a detector (UV or FID). At periodic intervals aliquots of the extract can be injected into the SFC column for analysis.

It is clear that static SFE-CSFC with aliquot sampling is a good technique for basic qualitative investigation and for measuring fundamental parameters such as partition coefficients of solutes in supercritical fluids. However, it is

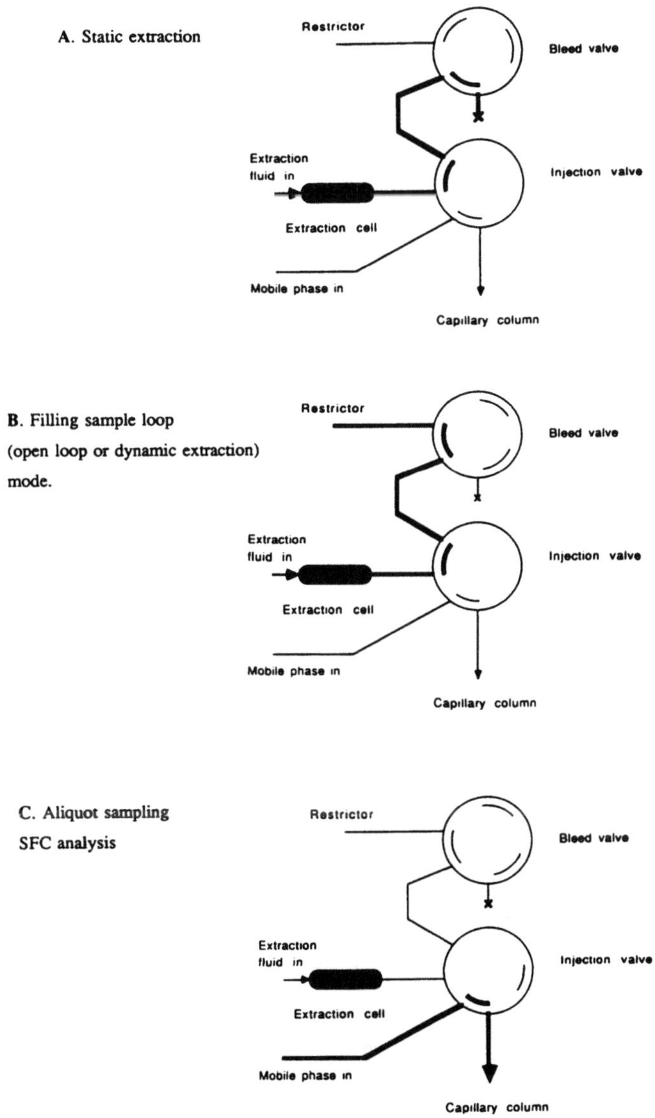

Figure 4.2 Configurations for static SFE-CSFC with aliquot sampling (reprinted with permission from reference 24).

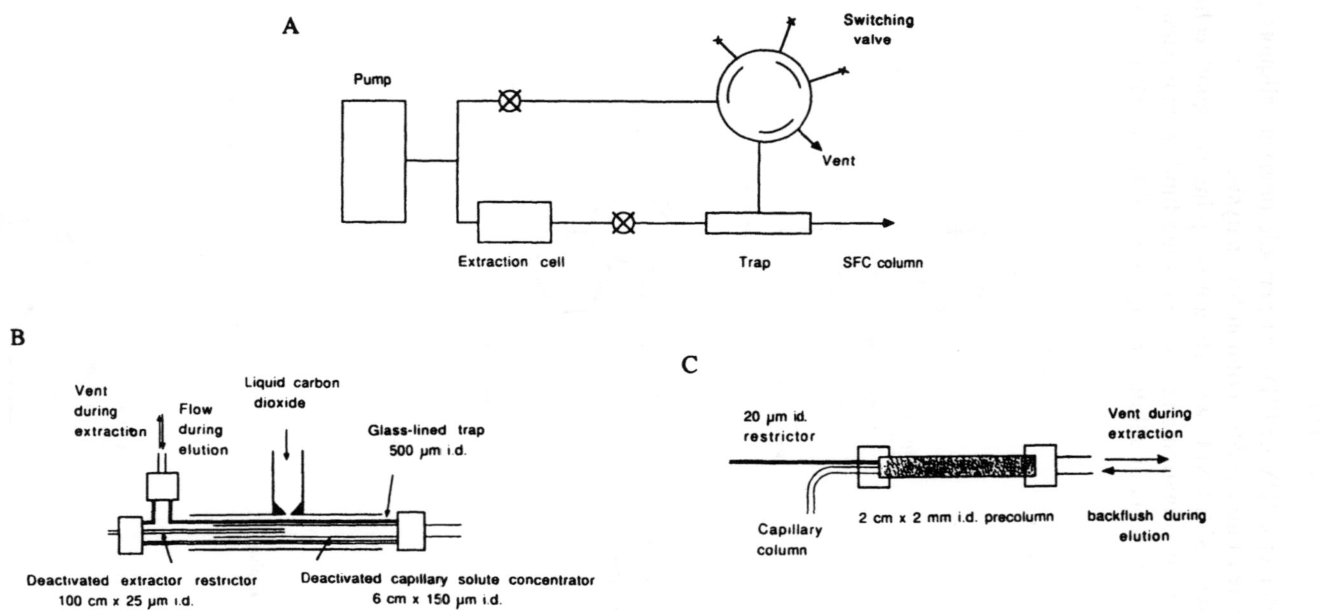

Figure 4.3 Dynamic SFE-CSFC with analyte accumulation. (A) Instrument configuration, (B) sorbent trap, (C) cryogenically cooled tee (reprinted with permission from references 36 and 43).

limited in that it is not suited to quantitative or trace analysis where analytes in the whole extract must be accumulated prior to chromatographic analysis.

4.4.2 Dynamic extraction with analyte accumulation

In contrast to static extraction with aliquot sampling, dynamic SFE-CSFC operates principally by continuously exposing the analytes to a fresh stream of supercritical fluid (Figure 4.2B) and accumulating the extracted components in a trap of some kind [24]. After extraction, the trapped analytes are transferred into the SFC column for analysis. The major advantages of dynamic SFE-SFC are that it is much more rapid than static SFE-SFC and that trace analysis can be performed. As shown in Figure 4.3, a high pressure syringe pump supplies the extraction cell with fluid. The outlet of the cell is connected to a capillary flow restrictor which is connected to an accumulating trapping system; this may take the form of a packed sorbent trap. However, in many cases an uncoated fused silica capillary tube housed in a low dead volume tee (cryogenically cooled) is used.

During extraction the tee is vented to the atmosphere, and the extract is concentrated within the uncoated fused silica retention gap or transfer line. After extraction is completed, the valve is switched and supercritical fluid is introduced into the side arm of the tee to transfer the analytes onto the SFC column. If uncoated fused silica tubing is used for the retention gap, the analytes will, in theory, be unretained during the transfer. Once they reach the stationary phase film of the SFC column they become concentrated as a narrow band due to differences in migration rates of the analytes in the two sections of the tubing. If such phase ratio focusing occurs successfully, then good chromatographic efficiency is observed during the separation [40]. Obviously, focusing is improved if the analytes have a greater concentration in the stationary phase when the mobile phase is introduced into the transfer region. This can be achieved with the mobile phase at low density in order to reduce the solubility of compounds in the fluid.

Other more complicated systems have been reported using on–off and multiport switching valves to allow continuous extraction or to permit the extraction cell to be vented during simultaneous chromatographic analysis [14, 41]. The 'plumbing' of such a system can be constructed to any specific requirement. Two different examples are shown in Figures 4.4 and 4.5. When incorporating switching valves into an SFE-CSFC system, attention should be paid to the pressure and temperature ratings of the valves, as leaks may develop if they are continuously used at high pressures (> 400 atm) and temperatures above 80°C. Further, the whole system (the pump, extraction cell, connecting tubing and all fittings) should be able to withstand high pressures and should be rated at least 1.5 times the maximum intended working pressure [42]. Attention should also be paid to the nature of the supercritical fluid used, particularly with respect to explosion hazards and the corrosion of equipment.

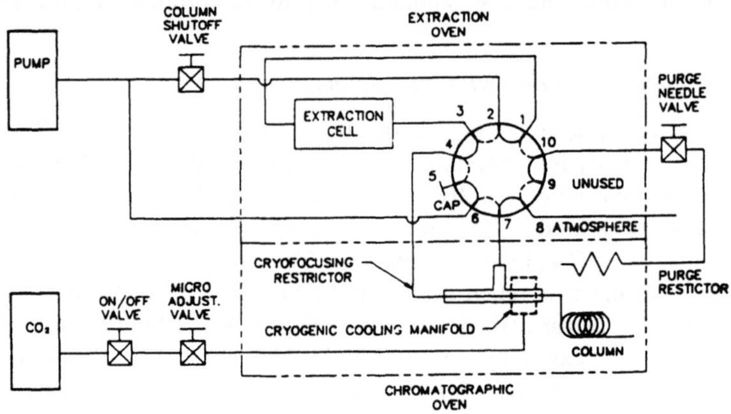

Figure 4.4 A schematic representation of the Lee Scientific 602 D SFE-CSFC apparatus. Flow is shown in the extract position (reprinted with permission from Lee Scientific Corporation).

When considering extraction vessels for SFE-CSFC, it is important that they are easy to open and close as well as being leak tight and properly rated for high pressures. The type of cell used will depend on the sample matrix (cells for liquid samples will be dealt with in a later section). In general, HPLC guard cartridges work particularly well for most solid matrices as the end caps are easily screwed on or off by hand. Frits (1–2 μm) are incorporated into the

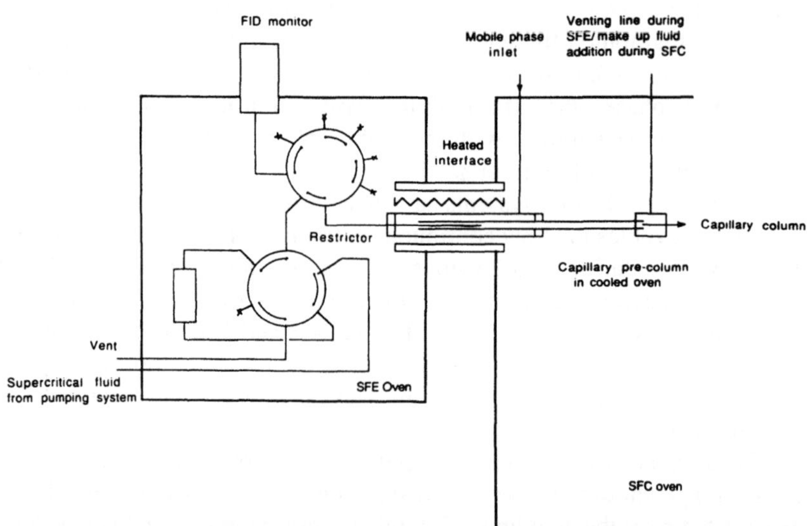

Figure 4.5 On-line SFE-CSFC system incorporating heated interface housing the SFE restrictor and cooled capillary pre-column for extract trapping.

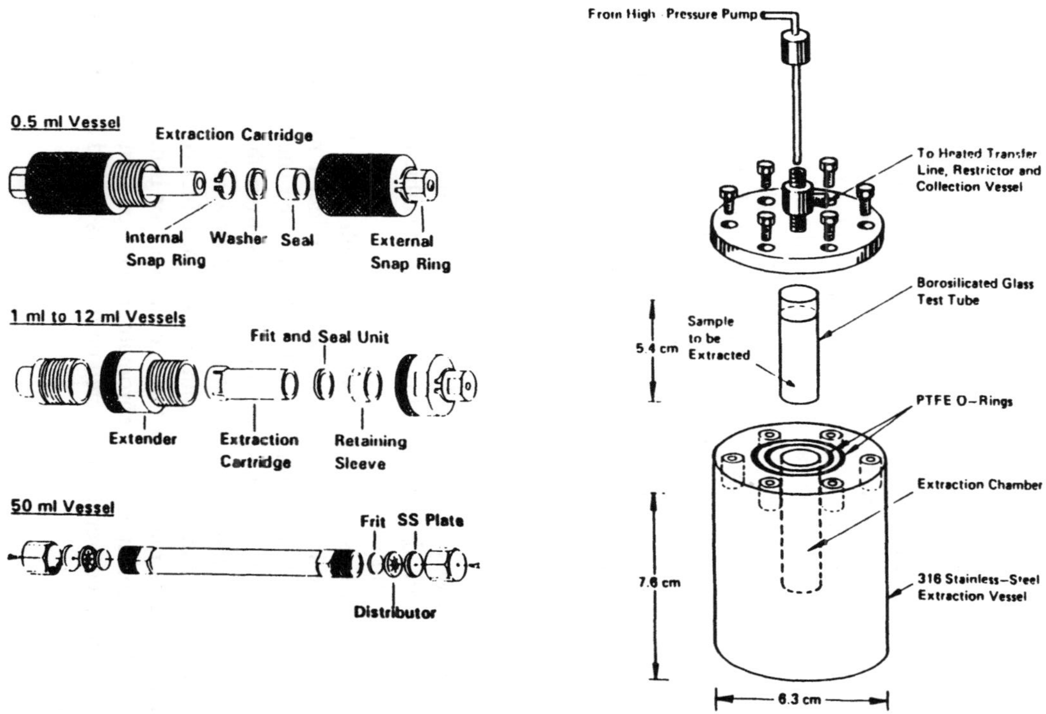

Figure 4.6 Typical extraction cells used in analytical supercritical fluid extraction.

endcaps to ensure that the sample matrix remains within the cell. Typical extraction cells used in analytical SFE [36, 43] are shown in Figure 4.6. Cells should be designed so that good contact of the supercritical fluid with the matrix occurs, and to this end it can be advantagous to mix the sample with inert silicaceous materials such as glass wool or pelletized celite.

With respect to dynamic SFE-CSFC with analyte trapping or accumulation, as in Figures 4.3–4.5, there are two important factors to consider: firstly, the total volume of extraction fluid that passes through the cell and secondly the flow rate of expanded extraction fluid passing through the trap.

Satisfactory recovery of solutes is dependent on analyte solubility and the volume of fluid with which the matrix comes into contact. Obviously, the greater the number of cell volumes of fluid passing through the cell per unit of time, the better the extraction efficiency. For example, it would be pointless to use a large extraction cell where only half a cell volume of fluid passed through the cell in the prescribed time and then try to quantify the amount of analyte in the sample. The aim is therefore to keep the volume of the extraction cell as small as possible by increasing the sample volume so that extraction can be carried out as quickly and efficiently as possible without affecting the analyte trapping procedure. With a knowledge of the cell geometry and the linear velocity (u) through the cell, the total number of cell volumes per unit time that pass through the extraction cell can be calculated from:

$$u = \frac{F_a \delta_c}{A_c \delta_a}$$

where F_a is the gaseous flow rate under ambient conditions, δ_c is the fluid density within the cell, δ_a is the fluid density under ambient conditions and A_c is the cross-sectional area of the extraction cell.

Since analytical SFE is most often performed with fluids that decompress to gases at ambient conditions (such as carbon dioxide, 1 ml/min of which produces a gas flow of approximately 500 ml/min), the success of trapping depends on the success of recovering the analytes from the expanded gas. Hawthorne *et al.* [27] have shown that expanded flow rates (in most cases expanded CO_2) greater than 300 ml/min reduce the effectiveness of the trap so that volatile solutes may be lost through the vent or band spreading may occur; both effects result in poor chromatographic performance. Thus, for quantitative results, recovery of extracted components should be performed at lower flow rates.

4.5 Trapping procedures

There are several methods of trapping extracted components from dynamic SFE in preparation for on-line capillary SFC analysis. The first reports of Raynor *et al.* [44] and Anton *et al.* [19] utilized the micro-injection valve of the SFC itself. Extracted components were deposited inside the internal volume rotor or sample loop during the dynamic extraction, and the valve was then switched to transfer them into the SFC column for separation. More recent studies have focused on the use of uncoated or coated fused silica transfer lines as retention gaps or retaining pre-columns, and on sorbent traps packed with HPLC materials.

4.5.1 Trapping in uncoated fused silica retention gaps

As was indicated in the previous section, the flow rate of the expanding extraction fluid and the temperature at which analytes are trapped, markedly affect the recoveries obtained when uncoated fused silica tubing is used for accumulating extracted components. In many systems, linear extraction restrictors are used, since they provide the correct flow rate range for on-line coupling with capillary SFC. They also tend not to plug as quickly as other restrictors when used for SFE. The length and internal diameter of the capillary restrictor tubing and the pressure at which the extraction is performed should therefore be considered to obtain suitable flow rates.

Restrictors with internal diameters greater than 30 μm result in higher extraction efficiencies, but lower recoveries and significant band broadening of more volatile components. However, restrictors with internal diameters less than 15 μm do not allow sufficient flow for efficient extractions over a short period of time, but yield good chromatographic peak shapes. As a rough guide, the gaseous flow rates from 15 cm lengths of 15, 20, 25 and 30 μm restrictors at a pump pressure of 300 atm are approximately 80, 150, 240 and 300 ml/min, respectively. A good compromise therefore is to use a restrictor with a flow rate of 100–200 ml/min. Lengths of capillary tubing of 20 or 25 μm diameter are suitable for most needs.

The trapping efficiency and subsequent chromatographic peak shape are also strongly dependent on the temperature at which analytes are accumulated. This is clearly shown by Figure 4.7 which reports the recoveries of extracted *n*-alkanes as a function of carbon number at different cryotrapping temperatures [22]. Too high a temperature yielded good recoveries and peak shapes for less volatile materials, but gave increasingly broad peaks and lower recoveries for more volatile analytes (earlier eluting components). The trap should therefore be cooled in the region of – 40°C to – 60°C. One problem with cooling the trap (if the restrictor is contained within the cooled zone Figures 4.3 and 4.4) is that too low a cryofocusing temperature may result in restrictor plugging, or components, such as water, freezing in the restrictor. This reduces the rate of extraction and makes it difficult to reproduce analyses. It may therefore be necessary to keep the restrictor hot and deposit the analytes in the transfer line held in a cryogenically cooled oven as shown in Figure 4.5.

System inertness is another important consideration. Deactivated fused silica tubing should be used for restrictors and transfer lines to reduce the possi-

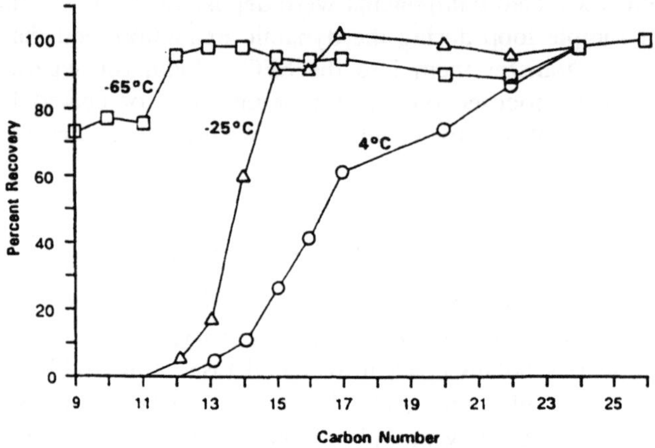

Figure 4.7 Recoveries of extracted *n*-alkanes as a function of carbon number at different cryotrapping temperatures (reprinted with permission from reference 22).

bility of adsorption of polar analytes onto active surfaces within the extraction system.

4.5.2 Trapping on coated fused silica retaining pre-columns

An alternative to the cryotrapping method is the use of a coated fused silica retaining pre-column for concentrating extracted solutes. Although trapping

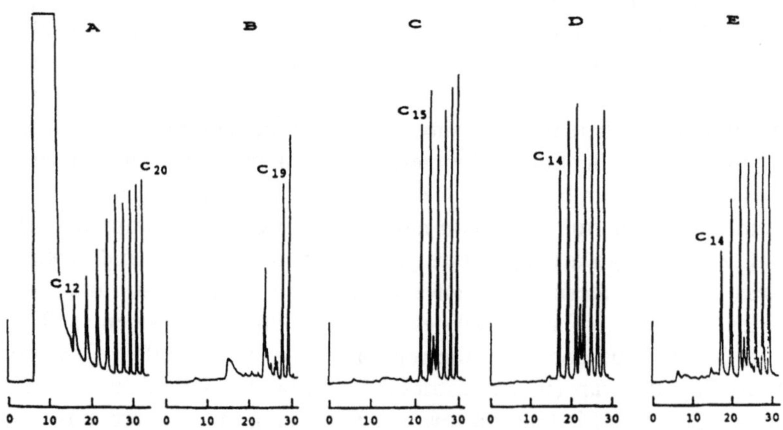

Figure 4.8 Comparison of sample recoveries using various traps in an SFE-SFC system. Chromatogram A was obtained by direct injection. Trap: uncoated capillary (B), capillary with a 0.1 μm film (C), capillary with 0.25 μm film (D), B plus packed column (E). Sample: *n*-paraffins (C_{12}–C_{20}), 0.5 μm of 100 ppm solution. Extracted and trapped at room temperature. Column: 10 m × 0.1 mm i.d. octyl, 0.5 μm. Column temperature: 90°C. Column pressure: 80 atm for A and 100 atm for the others for 10 min, then programmed at a rate of 5 atm/min. Detection: FID (reprinted with permission from reference 5).

techniques with such columns are continuously developing, Daimon *et al.* [5] have recently reported results comparing the recoveries of C12 to C20 alkanes at room temperature using an uncoated capillary, a 3 m × 100 μm i.d. capillary with 0.1 μm OV-1 film and a 3 m × 100 μm capillary with a 0.25 μm OV-1 film. Figure 4.8 summarizes the results obtained. Clearly there is band broadening and solute loss up to C18 with an uncoated capillary. However, if a retaining pre-column is used these compounds can be trapped effectively with maintenance of chromatographic peak shape even at room temperature.

4.5.3 Trapping on sorbent traps

Sorbents may also be used as an effective method of trapping after SFE [5]. This entails the use of short traps (usually 2 cm in length) packed with organic sorbents such as Tenax-GC, Carbotrap or with HPLC packing materials. It is important when using such a system that breakthrough of the analytes from the sorbent does not occur and also that the desorption behaviour is suitable for on-line chromatographic analysis. Desorption is performed by increasing the trap temperature or by using the supercritical fluid to desorb the sample. Figure 4.8(E) shows the recovery [5] of C12 to C20 *n*-alkanes from a 2 cm × 0.25 mm trap packed with ODS-300-5 HPLC material in on-line SFE-CSFC.

Munder *et al.* [4] have recently reported the analysis of explosives by SFE-CSFC using a 2 cm × 2 mm i.d. trap packed with aminopropyl bonded silica material (Alltech). Explosive residues extracted from soils were accumulated in the sorbent trap and later desorbed by backflushing the components with CO_2 onto the analytical column. It was found that the desorption process for each specific component depended on the solvating power of the CO_2 and that not all compounds were desorbed under all conditions. Hence, the desorption process is density and temperature dependent. In this regard, supercritical nitrous oxide has been found to be more effective than supercritical carbon dioxide in removing solutes from adsorbents. In some cases this may be useful, since selective desorption could be carried out for sample clean-up or fractionation procedures. This has already been demonstrated off-line by Campbell and Lee [45] who were able to fractionate complex coal-derived mixtures using silica adsorbents prior to GC analysis. Schaeffer *et al.* [46] have also successfully isolated an alkaloid on a cation exchange resin from a supercritical CO_2/ethanol fluid.

4.6 Optimization of conditions for SFE-CSFC

A number of parameters must be optimized for successful analysis by coupled SFE-CSFC. Principal among these are the conditions for quantitative extraction. This should begin with a determination of the supercritical fluid extractability of the analyte(s) from the non-sorptive matrices (filter paper, etc.) to assess the appropriate solvent, density and temperature conditions. Trial runs on spiked samples then allow investigations of matrix–solute interactions; if necessary these may be overcome by a period of static extraction. The kinetics of extraction must then be determined in order to define the required extraction time. A full discussion of these considerations is presented in chapter 1.

Factors affecting the efficiency of intermediate trapping must then be addressed. The nature of the analyte is crucial, while the possible presence of co-extracted, interfering compounds demands either selectivity during extraction, or the trapping on an adsorbent from which selective desorption into the SFC column is possible. The sample size must be carefully chosen so that the capacity of the SFC column is not exceeded, and the extracting supercritical solvent must be of sufficient purity to avoid introduction of extraneous material into the column [7]. Finally, the conditions for efficient SFC analysis must be optimized, preferably off-line. Correct choice of column, temperature and pressure/density programme are vital. Compromises may be inevitable if the extracted analytes have a range of polarities.

4.7 Selected applications of SFE-CSFC

Table 4.2 lists selected applications of SFE-CSFC. Many of these are real problems in the area of fuels, foods and food contaminants, and polymer chemistry.

The simplified procedure of trapping the CO_2 SFE extract in the rotor groove of the cooled injection valve was applied by Raynor *et al.* [44] to 1 mg samples of coal pitch. Switching the contents of the loop by means of the CO_2 mobile phase then permitted the analysis by CSFC (Figure 4.9); fractions containing individual hydrocarbon peaks were collected on a potassium bromide disc for subsequent identification by Fourier transform infrared microspectrometry, a procedure that allowed isomer identification not possible by SFC-mass spectrometry. King [47] applied a commercial SFE-CSFC system to a variety of food related analyses. Thus, the triglycerides and fatty acids of single seeds were characterized to give information required during genetic engineering studies. The SFC chromatogram after SFE of < 1 mg of a cotton seed kernel at 200°C and 45°C, showed the C50–C54 triglycerides of cottonseed oil which were injected after cryofocussing in the tee cooled by external CO_2 (Figure 4.4). Figure 4.10 is the SFE chromatogram of the on-line SFE extract obtained by the same procedure from a single dried-fruit beetle (*Carcophilus hemipterus*) showing the high molecular weight wax components.

A similar technique was employed by Cotton *et al.* [6] to the quantitative extraction and separation of additives and oligomers from polypropylene and a number of other polymers. Quantitative extraction of additives from polypropylene was investigated at five different extraction pressures, at a constant flow rate and temperature. Below 50 atm, extraction was negligible; between

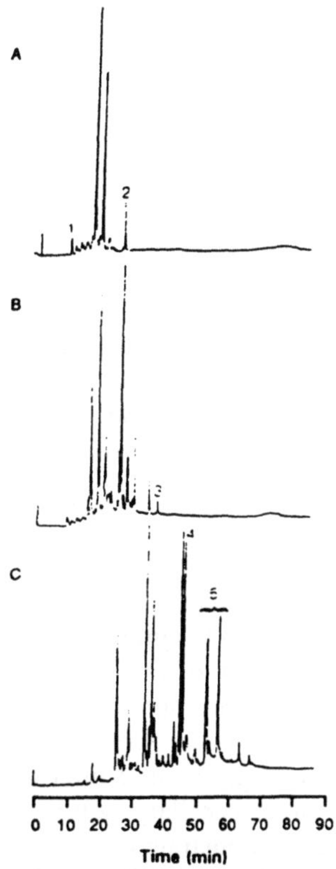

Figure 4.9 SFE-SFC chromatograms of a coal tar pitch extracted at (A) 70 atm, (B) 100 atm and (C) 200 atm. SFE conditions: CO_2; 43°C; 1 h. SFC conditions: 10 m × 50 μm i.d. open tubular column, poly (30% biphenyl)-methylsiloxane stationary phase; 110°C; density program from 0.25 g/ml to 0.74 g/ml at 0.006 g/ml per min after 20 min isobaric period; FID. Peak identification: (1) naphthalene, (2) phenanthrene, (3) pyrene, (4) chrysene, (5) benzopyrenes and benzofluoranthenes.

50 and 200 atm, Tinuvin 326 and 770 were extracted, along with small quantities of oligomers, although not in the ratio anticipated from their relative concentrations in the material. Higher pressures led to the extraction of all the additives present (Figure 4.11). The integrated peak areas correspond well with the actual concentrations (Table 4.3).

Table 4.3 Actual and relative SFE-SFC determined additive composition in commercial polypropylene [6]

Additive	Composition of polymer (%)	Relative composition	Relative peak area in extract
Tinuvin 770	0.4	1.33	1.30
Tinuvin 326	0.3	1	1
Irganox 1010	0.1	0.33	0.27
Irgafos 168	0.03	0.1	0.11

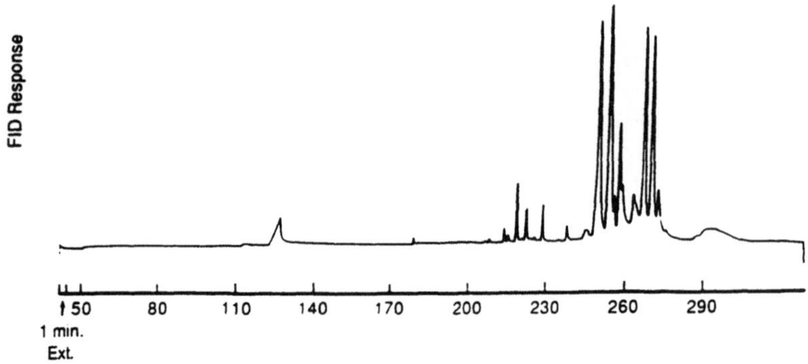

Figure 4.10 SFE-SFC separation of the on-line CO_2 extract from a single dried fruit beetle (live specimen). Extraction pressure = 200 atm; temperature = 45°C; time = 1 minute; extract cryofocused (reproduced with permission from reference 47).

The oligomer *tris*-(ethylene terephthalate), a cyclic trimer of poly(ethylene terephthalate) monomer, was identified [7] in the supercritical fluid extract of a commercially available PET film (Figure 4.12). The coefficient of variation for extraction and chromatography of the extracted cyclic trimer was determined for two extraction times to be approximately 8% as shown in Table 4.4.

Xie *et al.* [11] employed a cryogenic trap in which the supercritical fluid extract was decompressed through a small internal diameter linear restrictor and deposited in a deactivated capillary solute concentrator. The efficient operation of the system was demonstrated by the extraction (at 80°C and 400 atm), trapping (at – 40°C to – 50°C) and SFC analysis of the digitalis cardioglycoside ouabain, 3-[(6-deoxy-α-*l*-mannopyranosyl)oxyl]-1,5,11a,14,19-pentahydroxycard-20(22)-enolide (Figure 4.13). High recoveries of ouabain were deduced from biological assays of collected fractions.

Firearm propellants were characterized through their extractable organic constituents by SFE-CSFC by Munder *et al.* [4]. Detection limits of 100 pg were achieved by coupling three different detectors in-line (UV absorption, FID and ECD) and injecting the whole extract rather than a portion. Precipitation pre-concentration without cryotrapping was performed by expanding the fluid containing the analyte into a short sorbent packed column from which it was eluted by supercritical CO_2. Figure 4.14 illustrates the SFC traces which allowed the propellants present in trace gunshot residues remaining on expended cartridge cases to be identified.

Table 4.4 Coefficient of variation for the analysis of the cyclic trimer of PET by SFE-SFC [6]

Length of extraction (min)	Number of replications	Mass of PET extracted (mg)	Coefficient of variation (% RSD)
10	5	15.1 ± 0.2	8.2
25	18	15.2 ± 0.2	7.8

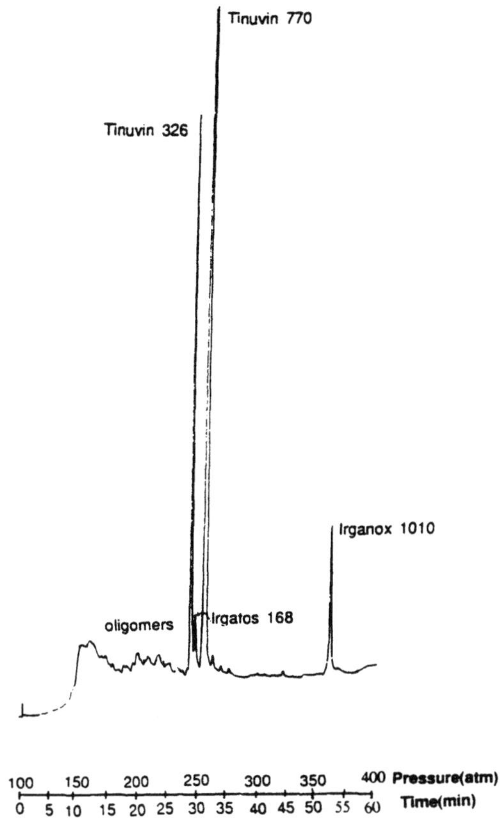

Figure 4.11 On-line SFE-CSFC of polypropylene pellets. Conditions: extraction with CO_2 at 70°C, 300 atm for 4 min. Chromatography with CO_2 at 100°C. Pressure programmed from 100 atm to 400 atm (at 5 atm/min). Column: 10 m × 50 μm i.d. SB-biphenyl; 6 mg introduced into the extraction cell.

Murugaverl and Voorhees [50] analysed trace amounts of pesticides in fats and oils by modifying the SFE-CSFC procedure to permit extraction and clean-up in one step. Extraction and clean-up were monitored by a photodiode array detector. The clean-up stage was achieved by selectively retaining extracted lipids on a packed column, while depositing analytes of interest in a cryogenic trap. In Figure 4.15, three pesticides are found in the SFC chromatogram of the first fraction obtained by extraction of ppm level spiked soybean oil; the lipids are found in the second and subsequent fractions.

4.8 Conclusions

On-line SFE-CSFC has shown itself over the last few years to be an invaluable and powerful coupled analytical procedure combining a sample preparation

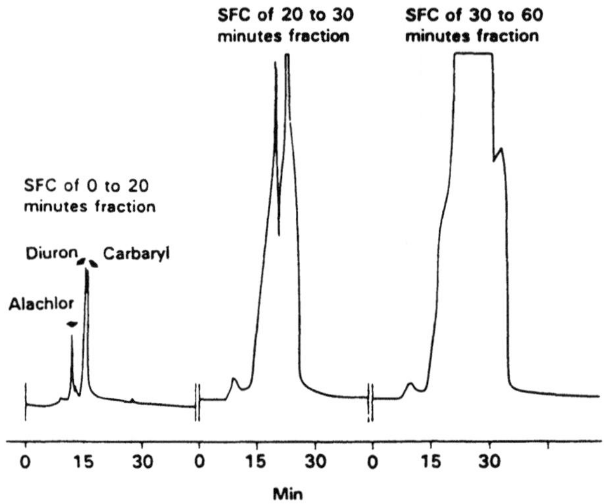

Figure 4.15 SFC chromatograms showing extraction of soybean oil spiked with alachlor (23 ppm), diuron (50 ppm) and carbaryl (42 ppm) (reproduced with permission from reference 50).

Acknowledgement

The authors would like to thank the Science and Engineering Council and ICI plc for financial support of N.J.C.

References

1. M. Saito, T. Hondo and M. Senda, *Multidimensional Chromatography — Techniques and Applications*, ed. H. J. Cortes, Chromatographic Science Series, Vol. 50, Marcel Dekker, New York (1990) pp. 331–357.
2. M. L. Lee and K. E. Markides, eds. *Analytical Supercritical Fluid Chromatography and Extraction*, Chromatography Conferences Inc., Provo, Ut (1990) pp. 344–362.
3. K. Sugiyama, M. Saito, T. Hondo and M. Senda, *J. Chromatogr.* **332** (1985) 107.
4. A. Munder, R. G. Christensen and S. A. Wise, *J. Microcolumn Sep.* **3** (1991) 127.
5. H. Daimon and Y. Hirata, *Chromatographia.* **32** (1991) 549.
6. N. J. Cotton, K. D. Bartle, A. A. Clifford, S. Ashraf, R. Moulder and C. J. Dowle, *J. High Resolut. Chromatogr.* **14** (1991) 164.
7. K. D. Bartle, T. Boddington, A. A. Clifford, N. J. Cotton and C. J. Dowle, *Anal. Chem.* **63** (1991) 2371.
8. R. C. Wiebolt, K. D. Kemptfert and D. C. Dalrymple, *Appl. Spectrosc.* **44** (1990) 1028.
9. Y. Hirata, F. Nakata and M. Horihata, *J. High Resolut. Chromatogr.* **11** (1988) 81.
10. E. D. Ramsey, J. R. Perkins, D. E. Games and J. R. Startin, *J. Chromatogr.* **464** (1989) 353.
11. Q. L. Xie, K. E. Markides and M. L. Lee, *J. Chromatogr. Sci.* **27** (1989) 365.
12. M. Saito, Y. Yamauchi, K. Inomata and W. Kottkamp, *J. Chromatogr. Sci.* **27** (1989) 79.
13. H. Englehardt and A. Gross, *J. High Resolut. Chromatogr.* **11** (1988) 38.
14. G. Mapelli, F. Pigozzo, M. W. Raynor and S. Trestianu, 13th Int. Symp. on Capillary Chromatography, ed. P. Sandra, Riva del Garda, Italy (1991) p. 1383.
15. D. Thiebaut, J. P. Chervet, R. W. Vannoort, G. J. DeJong, U. A. Th. Brinkman and R. W. Frei, *J. Chromatogr.* **477** (1989) 151.
16. J. Hedrick and L. T. Taylor, *Anal. Chem.* **61** (1989) 1986.
17. W. Gmuer, J. O. Bosset and E. Plattner, *J. Chromatogr.* **388** (1987) 335.

18. R. J. Skelton, C. C. Johnson and L. T. Taylor, *Chromatographia* **21** (1986) 3.
19. K. Anton, R. Menes and H. M. Widmer, *Chromatographia* **26** (1988) 221.
20. S. B. Hawthorne, D. J. Miller and M. S. Krieger, *J. Chromatogr. Sci.* **27** (1989) 347.
21. S. B. Hawthorne, M. S. Krieger and D. J. Miller, *Anal. Chem.* **60** (1988) 472.
22. M. R. Anderson, J. T. Swanson, N. L. Porter and B. E. Richter, *J. Chromatogr. Sci.* **27** (1989) 371.
23. K. R. Jahn and B. Wenclawiak, *Chromatographia* **26** (1988) 345.
24. J. R. Wheeler and M. E. McNally, *J. Chromatogr. Sci.* **27** (1989) 534.
25. F. I. Onuska and K. A. Terry, *J. High Resolut. Chromatogr.* **12** (1989) 527.
26. J. M. Levy, R. A. Cavalier, T. N. Bosch, A. M. Rynaski and W. E. Huhak, *J. Chromatogr. Sci.* **27** (1989) 341.
27. S. B. Hawthorne and D. J. Miller, *J. Chromatogr.* **403** (1987) 63.
28. S. B. Hawthorne and D. J. Miller, *J. Chromatogr. Sci.* **24** (1986) 258.
29. M. W. F. Nielen, J. T. Sanderson, R. W. Frei and U. A.Th. Brinkman, *J. Chromatogr.* **474** (1989) 388.
30. S. A. Liebmàn, E. J. Levy, S. Lurcott, S. O Niel, J. Guthrie, T. Ryan and S. Yocklovich, *J. Chromatogr. Sci.* **27** (1989) 118.
31. J. M. Levy and J. P. Guzowski, *Fresenius Z. Anal. Chem.* **330** (1989) 207.
32. J. M. Levy, J. P. Guzowski, and W. E. Huhak, *J. High Resolut. Chromatogr.* **10** (1987) 337.
33. S. B. Hawthrone, M. S. Krieger and D. J. Miller, *Anal. Chem.* **61** (1989) 736.
34. K. D. Bartle, A. A. Clifford, S. B. Hawthrone, J. L. Langenfield, D. J. Miller and R. Robinson, *J. Supercrit. Fluids* **3** (1990) 143.
35. S. B. Hawthorne, *Anal. Chem.* **62** (1990) 633A.
36. B. W. Wright, C. W. Wright, R. W. Gale and R. D. Smith, *Anal. Chem.* **59** (1987) 38.
37. J. W. King, *J. Chromatogr. Sci.* **27** (1989) 355.
38. K. S. Nam, S. Kapila, G. Pieczonka, T. E. Clevenger, A. F. Yanders, D. S. Viswanath and B. Mallu, *Proc. Int. Symp. on Supercritical Fluids* (1988) p. 743.
39. W. P. Jackson, K. E. Markides and M. L. Lee, *J. High Resolut. Chromatogr.* **9** (1986) 213.
40. G. Schomburg and W. Roeder, *J. High Resolut. Chromatogr.* **12** (1989) 218.
41. Lee Scientific Series 600-D Supercritical Fluid / Gas Chromatograph User's Manual.
42. M. Saito, T. Hondo and Y. Yamauchi, *Supercritical Fluid Chromatography*, ed. R. M. Smith Royal Society of Chemistry, Cambridge (1988).
43. M. L. Lee and K. E. Markides, eds., *Analytical Supercritical Fluid Chromatography and Extraction*, Chromatography Conferences Inc., Provo, Ut (1990) p. 349.
44. M. W. Raynor, I. L. Davis, K. D. Bartle, A. A. Clifford, A. Williams, J. M. Chalmers and B. W. Cook, *J. High Resolut. Chromatogr.* **11** (1988) 766.
45. R. M. Cambell and M. L. Lee, *Anal. Chem.* **58** (1986) 2247.
46. S. T. Schaeffer, L. H. Zalkov and A. S. Teja, *Ind. Eng. Chem. Res.* **28** (1989) 1017.
47. J. W. King, *J. Chromatogr. Sci.* **28** (1990) 9.
48. P. Manninen, Y. Holm, R. Hittlinen and M. L. Riekkola, *11th Int. Symp. on Capillary Chromatography*, eds. P. Sandra and G. Redant, Monterey CA (1990) p. 676.
49. F. Hofler and G. Alt, *11th Int. Symp.* on Capillary Chromatography, eds. P. Sandra and G. Redant Monterey CA (1990) p. 682.
50. B. Murugaverl and K. J. Voorhees, *J. Microcolumn Sep.* **3** (1991) 11.

5 Supercritical fluid extraction coupled to packed column supercritical fluid chromatography

I.G.M. ANDERSON

5.1 Introduction

Although the solvating ability of supercritical fluids was first recognized by Hannay and Horgarth working with supercritical ethanol in 1897 [1], supercritical fluids found little application as extraction media until the early 1960s when Zosel [2], working at the Max Planck Institute for Coal Research, introduced supercritical fluid extraction (SFE) as an industrial process, although Zhuze and co-workers had described an earlier process for the extraction of lanolin from wool grease in 1958 [3].

Supercritical fluids, such as supercritical carbon dioxide, were soon found to be highly efficient extraction media, chiefly on account of their high solvating power, their low viscosities (intermediate between a gas and a liquid) and their low surface tensions which enable their penetration deep into the extraction matrix.

SFE used in isolation is generally not selective enough to isolate specific solutes from the extraction matrix without further clean-up or resolution from co-extracted species prior to qualitative and quantitative analysis. Consequently, for analytical applications, SFE is usually used in conjunction with chromatographic techniques, to improve the overall selectivity of the process in isolating specific solutes.

The combined use of SFE with chromatographic techniques is quite recent. The first example is that of Stahl and Schiltz, who described the use of a micro extractor with thin-layer chromatography (TLC) in 1976 [4]. Since 1976, SFE has found increasing application with chromatographic techniques such as high performance liquid chromatography (HPLC) [5], gas chromatography (GC) [6,7] and supercritical fluid chromatography (SFC).

SFE combined with chromatography can be either 'off-line' or 'on-line'. In the off-line process supercritical fluid extraction takes place as a separate and isolated process to the chromatography: extracted solutes are trapped or collected, often in a suitable solvent for later injection onto GC, HPLC or SFC instrumentation. In the on-line process, SFE and chromatography are coupled to form an integrated process: following SFE the extracted species are passed directly to the chromatograph, usually via a trap or sample loop and a valve switching device.

Supercritical fluid chromatography, in which a supercritical fluid is used as the mobile phase, was introduced by Klesper in 1962 [8]. Advances in instrument technology, sophistication and computation over the next twenty years led to the development of increased on-line coupling of chromatographic (and other analytical) techniques. In the mid 1980s Sugiyama, Saito and Wada [9–11] and, separately, Skelton, Johnson and Taylor [12] introduced SFE which was directly coupled to SFC, combining the advantages of both techniques.

Supercritical fluid chromatography can be performed on either packed or capillary columns and although Sugiyama *et al.* [10] used a packed HPLC column in their initial work on directly coupled SFE-SFC to extract and resolve caffeine from coffee beans, capillary column SFC was more widely used until the late 1980s when there was, and continues to be, a renewed interest in packed column SFC.

This chapter reviews the reasons for this renewed interest in packed column SFC. Advantages and disadvantages of packed columns and their packing materials are compared with capillary columns, particularly for SFE coupled to packed column SFC. SFE coupled to packed column SFC is discussed with introductory sections on supercritical fluid mobile phase, supercritical fluid extraction and supercritical chromatography. Details of instrumental aspects of supercritical extraction coupled with packed column SFC are reviewed, highlighting some relevant applications to illustrate their use.

Instrumental aspects discussed are: back pressure regulators, extraction vessels, on-line solute trapping and collection devices, on-line solute introduction, columns, detectors and fraction collection. Specific details of commercially available pumps for supercritical fluid delivery and of commercial detectors used for SFC will not be given here since detailed information is available from the instrument manufacturers.

Following the review of instrumental components, applications of SFE coupled with packed column SFC are detailed and discussed followed by some thoughts on the future of the coupled technique.

5.2 Packed vs. capillary columns

Three types of column are used for SFC: capillary, packed capillary and packed. The physical differences between these three types of column, in terms of column diameter and length, film thickness or particle size, are summarized in Figure 5.1, together with listings of the most commonly used stationary phase materials for the columns. In this chapter, packed capillary columns are included with packed columns, under the general term 'packed columns'.

In their initial application of SFE directly coupled to SFC, Sugiyama *et al.* [10] used packed column SFC. However, the majority of workers in the field

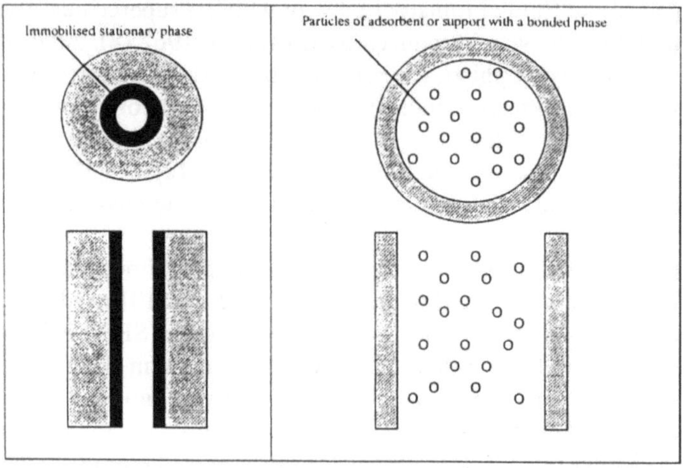

Nomenclature	CAPILLARY	PACKED CAPILLARY	PACKED
Internal Diameter	25 - 100 µm	0.25 - 1 mm	2 - 4.6 mm
Length	3 - 20 m	10 - 100 cm	10 - 25 cm
Film Thickness	0.05 - 1 µm		
Particle Size		3 - 10 µm	3 - 10 µm
Stationary Phase	Methylsilicon Phenyl Cyanopropyl Biphenyl PEG Chiral Phases Liquid Phases	Silica C_{18} - C_8 Diol Amino Cyano Chiral Phases	

Figure 5.1 Types of column used for supercritical fluid chromatography. (Redrawn with permission from Carlo Erba.)

preferred to use capillary columns until quite recently, partially for two historical reasons. First, SFC was mainly developed from high pressure gas chromatography technology [8] and many of the early workers had originated from a gas chromatography background, and second, open tubular capillary columns had been introduced by Novotny [13] in 1981, coincidental with Hewlett–Packard's launch of commercial SFC instrumentation at that year's Pittsburg Conference.

Apart from these historical reasons, there were two major practical factors which influenced th early preference for the use of capillary over packed columns. These were the problems of pressure drop over the length of the column, emphasized by Novotny [13], and of the inertness of surface coatings used for capillary columns when compared to packed column SFC, where the majority of packing materials are based on bonded silicas with high surface activity due to residual silanol groups. Both these problems will be discussed.

Differences, advantages and disadvantages of packed and capillary columns and between packed and capillary SFC have been reviewed by a number of authors. For example, Wheeler and McNally [14] compared packed and capillary SFC with HPLC using representative herbicides and pesticides to evaluate the use of the supercritical fluid techniques as an alternative to HPLC, and Schoenmakers [15] used a mathematical approach to compare packed columns and open tubular capillary columns for SFC in terms of column efficiencies, column pressure drop, influence of film thickness, speed of analysis and the maximum number of theoretical plates available for each of the column types. More recent papers where packed and capillary columns have been compared include those of Petersen [16], which deals with packed columns in SFC and Schoenmakers *et al.* [17], in which a number of stationary phases were evaluated and compared for packed column SFC, principally on the basis of packing material inertness. Griebrokk *et al.* [18] and Taylor and Chang [19] have discussed the current use and development of packed columns in SFC.

These papers identify the major advantages and disadvantages of packed and capillary columns for SFC as outlined in the following sections.

5.2.1 Efficiency

Approximately identical numbers of theoretical plates are obtained per unit length of packed and capillary column when the particle size of the packed column is equivalent to the internal diameter of the capillary column [20]. An average internal diameter of a capillary column is 50 μm and as particle sizes for packed columns are in the range 3–10 μm, so the number of theoretical plates per unit length of packed column is higher than that for a capillary column. However, set against this is the large difference in permeability between packed and capillary columns, since permeability is obviously greater for an open tubular column than for a packed column.

Permeability influences the pressure drop along the length of the column and as the pressure drop increases, the supercritical fluid mobile phase density gradient increases. Density is directly linked to the capacity factor (k^1) and therefore a capacity factor gradient exists having a low value of (k^1) at the column inlet and a higher value at the column outlet. Under certain circumstances this negative capacity factor gradient can lead to the practical result of a decrease in chromatographic resolution with column length. For a given pressure drop over the column, capillary column SFC will allow 100 to 500 times more theoretical plates than packed column SFC with 5 or 10 μm diameter particles [15].

For difficult separations, it is generally better to use the resolving power of capillary SFC. However, for most practical purposes short packed SFC columns will be able to provide enough theoretical plates for most separations when other factors, such as advantages of packed column selectively, are taken into account.

5.2.2 Selectivity

The phase ratio, V_s/V_m, is higher for packed than capillary columns, which means that interactions between the stationary phase and the sample solutes are of more significance for packed than capillary columns. This provides a major advantage for packed column SFC over capillary columns in terms of their selectivity towards specific solutes, since advantage can be taken of the wide range of packing material selectivities commercially available for conventional HPLC columns, compared with the rather narrow range of surface coatings available for capillary columns. Additionally, Pirkle-type stationary phases, as well as cyclodextrins and bonded chiral diamines, are generally confined to packed columns where they may be used to separate optically active enantiomers such as amides and pairs of optically active phosphine oxides.

A major problem of packed columns, especially those based on bonded silica has been the surface activity of the packing materials, principally due to residual silanol groups as a consequence of incomplete or patchy coverage of the porous silica particles with the bonded phase material. On chromatography with polar compounds this may cause asymmetric peak shape and sometimes severe tailing due to strong hydrogen bonding, ionic and dipole interactions with the stationary phase, particularly with acidic and basic compounds, which can result in their partial retention on the chromatographic column through irreversible absorption.

This surface activity of early packing materials for packed columns was the second and perhaps most serious reason why early SFC workers preferred capillary columns, since the internal surface area of capillary columns is covered with sufficient film thickness of stationary phase surface coating to cover the silanol groups present, rendering the columns inert. For this reason, capillary column SFC has been principally employed for the analysis of polar solutes. Packed column SFC is used more commonly for non-polar solutes such as polynuclear aromatic hydrocarbons since elution of polar materials from the commonly used non-polar C18 bonded phase silicas are difficult with supercritical carbon dioxide, due to its lack of polarity and the interaction of polar solutes with the surface activity of the silica based materials.

In recent years there has been much effort, and a certain amount of success, in reducing the surface activity of silica based packing materials. This, combined with the introduction of new packing materials, the introduction of smaller diameter and more uniform particles, together with the ability to pack narrower fused silica columns, has primarily resulted in the increased interest in packed column SFC [18]. Also, packed column SFC has a number of significant benefits over capillary SFC, particularly for SFE coupled with SFC and it could be argued that part of the renewed interest in the use of packed SFC columns comes directly from the benefits of SFE coupled to SFC.

Partial reduction in surface activity has been achieved by 'end-capping' bonded silicas and by other surface treatments aimed at shielding residual silanols. Hydrosilyloxane polymers have been used to shield residual silanol groups by their chemical bonding to porous silica particles using a coating technique similar to that developed for capillary GC [21]. In this technique a less active surface is produced by using a cross-linking agent to combine the polysiloxane chains which mask the silanol groups.

Several non-silica based packing materials have been used which overcome the silica based packed column limitation of surface activity. Better results were obtained for the separation of iso-octylphenoxy polyethoxy ethanol oligomers from Triton X-100 surfactant by SFC on a reversed-phase PRP polymer column than on a C18 bonded silica column [22] although there may be some problems with using polymer resin columns under the high pressures required for SFC [19].

Porous graphitic carbon developed by Knox *et al.* [23] has potential as an inert packing material for packed column SFC since it provides an inert hydrophobic medium with little or no surface activity. Schoenmakers *et al.* [17] looked at a number of different stationary phases for packed column SFC in terms of their surface activities. These included modified and unmodified porous graphitic carbon and modified and unmodified poly(styrene-divinylbenzene) copolymers as well as chemically modified silicas and polysiloxane-coated silicas, discussed earlier. The compatibility of the phases with the supercritical carbon dioxide mobile phase, and a number of test solutes of different polarities, was established from qualitative examination of their eluted peak shapes. Polysiloxane-coated silica packing materials performed better than C18 bonded phase silica for certain solutes but some peak shapes and efficiencies were still poor. Of the packing materials tested under a uniform set of conditions, the unmodified carbon column proved to be most stable. By experimentally coating poly(ethylene glycol) onto porous graphitic carbon a 'carbonwax' column was prepared which provided the best, non-active packing material for the SFC of polar solutes.

Supercritical carbon dioxide is the most commonly used mobile phase for SFC on account of the ease with which supercriticality may be accomplished ($T_c = 31°C$, $P_c = 73$ bar), its availability in a pure form, non-toxic nature, cheapness and compatibility with a number of chromatographic detectors. However, it provides an essentially non-polar mobile phase which has poor eluting characteristics for polar materials in the most common stationary phases used for packed column SFC.

Although recent developments in packing materials have resulted in increased inertness, there is still some way to go in terms of reducing surface activity and its interaction with polar solutes. Because of the latter, polar materials tend to chromatograph poorly and have relatively long retention times with packed columns and unless a more polar supercritical fluid mobile phase is used, some other means must be found to deactivate the packing material. Of the supercritical fluids available the polar supercritical ammonia may be a possible choice as an alternative to the non-polar carbon dioxide.

However, it would be obnoxious to use and a better way of improving the polarity of the mobile phase is to add a small amount of a polar organic solvent, commonly 2% or less of methanol, to the supercritical carbon dioxide as a 'modifier' or 'entrainer'. The addition of polar modifier increases the mobile phase polarity and also deactivates surface silanol groups resulting in better peak shapes and reduced retention time for polar solutes. The addition of a small amount of modifier also increases the critical temperature and density of the mixture. This increases the solvation power of the mobile phase, further reducing the retention of polar solutes. Modification of solute retention times may also be achieved by applying a modifier gradient with packed column SFC using a binary pumping system.

Columns for capillary SFC are more inert than packed columns, but less selectivity towards particular solutes is available because there is a smaller range of commercial stationary phase materials. The inertness of capillary columns provides better chromatography for polar compounds without the need for polar modifiers, although these can be added to the supercritical fluid at source. However, GC-type detectors used with SFC capillary columns are not compatible with mixed carbon dioxide modifier phases and this tends to restrict the use of polar modifiers to packed columns where ultraviolet (UV) detectors, compatible with both supercritical carbon dioxide and most polar modifiers, are mainly used.

5.2.3 Sample capacity

Because of their higher phase ratio, packed columns also have a greater sample capacity than capillary columns. This means that greater amounts of material can be added directly to the column without the need of sample splitting, commonly required for capillary SFC, which simplifies sample introduction by injection or sample valve. Higher sample load onto a packed column allows smaller concentrations of materials to be detected and as a consequence detection limits are often better from packed than capillary columns, particularly where small microbore (1–2 mm i.d.) columns are used. However, sensitivity is also dependent on detector characteristics.

The ability to load significantly more material onto a packed column is an important consideration and a major advantage of packed column SFC directly linked with SFE since the process of solute transfer from SFE is easier without sample splitting and ideal when an initially large sample is required for SFC followed by component collection or fractionation.

The large sample capacity of packed columns is also an important consideration if transfer to semi-preparative or preparative SFC is being considered.

5.2.4 Detectors

The low mass flow from capillary columns means they are amenable for use with the gas detectors used for GC, such as flame ionisation, flame photometric and the nitrogen phosphorous specific detectors. Capillary columns are also directly compatible with mass spectrometric (MS) and Fourier transform infrared (FTIR) detectors. With greater supercritical mobile phase flow, packed columns are more compatible with liquid detectors traditionally used for HPLC, such as the fixed or variable wavelength ultraviolet (UV) detectors, and multiwavelength or photo-diode array detectors, which are commonly used for packed column SFC.

Packed capillary columns have the advantage of being compatible with both gas and liquid detectors used for SFC, and are useful in cases where SFE coupled with SFC requires the introduction of relatively large samples to the SFC column followed by a FTIR or MS finish.

Overall the applicability of such a large range of both gas and liquid detectors is seen as one of the main attractions of SFC [23].

5.2.5 *Analysis times*

Schoenmakers has developed simple mathematical expressions to compare the speed of analysis in an open capillary with a packed SFC column and concluded that in terms of speed of analysis, capillary SFC columns could not compete with packed column SFC unless the column diameters were greatly reduced [15]. Hence, for speed of analysis packed SFC columns are better and because of this are often more suited to method development [17].

5.2.6 *Restrictors*

Capillary column SFC generally uses capillary restrictors to control supercritical fluid pressure. These are prepared from lengths of fused silica tubing and, apart from the difficulties experienced in obtaining reproducible restrictors, they tend to become clogged during SFC due the deposition of ice from moisture in (most) samples which may become frozen at the point of exit from the capillary restrictor due to the cooling effect of the evaporating carbon dioxide. Much more robust back pressure regulators, which do not clog and are able to maintain a more reproducible pressure in the SFC system, are used for packed column SFC systems. This important consideration helps to make SFE coupled to packed column SFC more stable and reproducible than SFE coupled with capillary SFC.

5.2.7 *Temperature*

As capillary SFC columns were developed from GC stationary phases they can be used at higher temperatures than packed columns, which were originally developed for use at the ambient or low temperatures used for HPLC. An increase in temperature decreases the fluid density and solvating ability of the mobile phase which in turn will increase the retention time for some compounds. However, it is generally the case that an increase in temperature in CSFC decreases retention times due to the increased volatility of volatile solutes. Therefore retention time behaviour in CSFC can be complex with some compounds increasing and some decreasing in retention with increasing temperature.

At increased temperatures with silica packing materials there may be some problems with the use of modified supercritical carbon dioxide, due to its high solvating power dissolving the silica backbone and/or bonded phases. The effects of high temperatures on conventional HPLC bonded phase materials and polymer coated bonded phase packings especially developed for SFC have been studied by Taylor and Chang [19] who compared a number of commercially available conventional C18, aminopropyl and silica packings using retention times and peak shapes from a number of test solutes. Possible loss of stationary phase during heat treatments was examined by thermal gravimetric analysis linked to mass spectrometry (TGA-MS). After heat treatment, retention time was reduced and peak shape improved for C18 material. TGA-MS revealed that only water had been lost from the column at temperatures up to 150°C. However, aminopropyl columns treated in the same way gave increased retention times and poorer chromatographic results. TGA-MS revealed loss of water and some aminopropyl stationary phase which illustrates that the loss of stationary phase from conventional HPLC packing materials is dependent on the exact nature of the bonded phase and, of course, the conditions used.

Taylor and Chang [19] also compared a number of polymer coated hydrosiloxane packing materials, including polymer coated cyanopropyl and phenyl, specially developed for SFC with conventional bonded phase packing materials and concluded that the polymer coated materials had less surface activity and were more thermally stable than the conventional phases.

In summary, the cross-linked stationary phases used for capillary SFC columns are generally stable at elevated temperatures under SFC conditions. Although conventional HPLC packing materials employed for packed column SFC are generally unstable at elevated temperatures, the newer polymer coated packing materials exhibit much greater thermal stability and consequently both capillary and packed column SFC can make use of elevated temperatures to reduce retention times of solutes, provided that the stability of the packing material is checked before use.

Apart from the major differences and similarities between packed and capillary SFC columns discussed above, Wheeler and McNally [4] found packed column SFC easier to use, having no need for injection splitters, and that the back pressure regulators used with packed columns were easier to use than the capillary restrictors used with capillary columns. Furthermore, better injection to injection reproducibility with less sample carry-over was found for packed column SFC.

5.3 SFE coupled to packed column SFC

The major differences between packed and capillary columns for SFC have been reviewed and where applicable comments have been made on the benefits of packed columns for SFE-SFC. We will now deal with equipment and techniques used for the direct 'on-line' coupling of SFE with SFC.

5.3.1 Supercritical fluid mobile phase

The most commonly used supercritical fluid for SFE, SFC and for SFE coupled with SFC is carbon dioxide on account of its ease of supercriticality ($T_c = 31°C$, $P_c = 73$ atmospheres), its availability at high purity, inertness and lack of toxicity and flammability. At typical working pressures (80–600 atmospheres) supercritical fluid carbon dioxide is an excellent solvent for non-polar solutes such as alkanes and terpenes and a reasonably good solvent for moderately polar compounds such as polynuclear aromatic hydrocarbons (PAHs), polychlorinated biphenyls (PCBs), aldehydes, esters, alcohols, organochlorine pesticides and fats, but is less useful for polar compounds [24]. Polar compounds, however, may be successfully extracted by the addition of a small amount of polar modifier to the sample or to the supercritical carbon dioxide mobile phase. As a rule of thumb solutes that can be determined by conventional gas chromatographic techniques can be quantitatively extracted using supercritical carbon dioxide [24].

A number of other substances can be used as supercritical fluids, such as the more polar nitrous oxide, ammonia, methanol, ethane and ethylene. However, these other compounds are more difficult to use in terms of handling and tend to be used only for more specialist applications, where supercritical carbon dioxide cannot be used alone or with an appropriate modifier.

5.3.2 Supercritical fluid extraction (SFE)

In its simplest form a supercritical extractor consists of a high pressure pump, similar to or adapted from pumps used for HPLC, an extraction vessel, a back pressure regulator and a separation vessel or trap where the extracted solutes of interest are collected. There are three basic types of SFE system characterized by the way in which the solutes are separated from the supercritical fluid extraction media used [25]. In the first type, solutes are separated from the extraction media based on pressure reduction which causes a solubility decrease. In the second type, a temperature change is used to bring about a decrease in solubility from the extraction media and in the third type the solutes are absorbed onto an appropriate adsorbate. Often a combination of the first and second is used, where after extraction the supercritical fluid is simply evaporated to leave the solutes of interest.

Supercritical fluid extraction can be either 'static' or 'dynamic'. In static SFE the sample and the supercritical fluid are introduced into the extraction vessel and left, under continued supercritical conditions, at a specific pressure and temperature while extraction takes place for a period of time after which the fluid is removed from the vessel and the solutes of interest collected. If extraction is not complete in the first 'pass', subsequent cycles of extraction with fresh supercritical fluid may be used. In dynamic SFE the sample in the extraction vessel is continually flushed or swept with supercritical fluid so that extraction is a continuous process during the extraction period. Using dynamic extraction the solutes of interest are either recovered, as before, by the evaporation of the supercritical fluid, or are trapped onto some suitable adsorbent media such as silica, tenax or activated charcoal. Further details of supercritical extraction and of collection and trapping methods used by workers in the field will be discussed later.

5.3.3 Supercritical fluid chromatography (SFC)

A packed column supercritical fluid chromatography (PSFC) system is very similar to a HPLC system except that there is a back pressure regulator, which pressurizes the mobile phase above its supercritical pressure, and a column heater is employed to keep the mobile phase above its critical temperature. The back pressure regulator maintains the system pressure through the detector for packed column SFC, unlike most cases for capillary SFC where use is made of a flame detector which does not operate under the high pressures used for SFC. Figure 5.2 shows a typical schematic for a packed column SFC system.

5.3.4 SFE coupled to packed column SFC (SFE-SFC)

SFE directly coupled to SFC provides an extremely powerful analytical tool. The efficient, fast and selective extraction capabilities of supercritical fluids

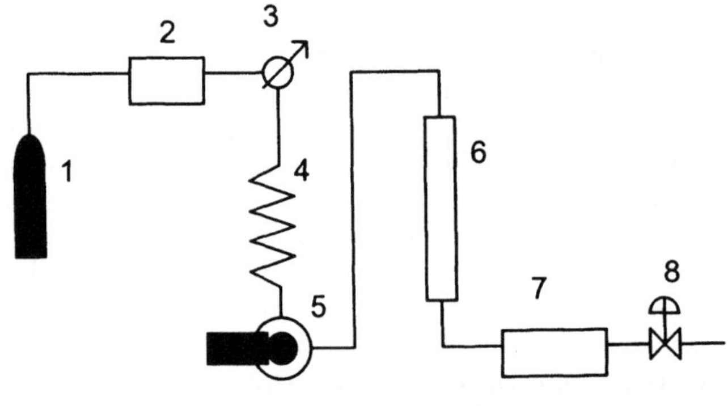

Figure 5.2 Schematic for packed column SFC: 1. gas cylinder; 2. pump; 3. pressure meter; 4. heat exchanger; 5. sample introduction valve; 6. separation column; 7. detector; 8. back pressure regulator.

allows the quantitative extraction and direct transfer of the selected solutes of interest to be accomplished to the column, often without the need for further sample treatment or clean-up. Extraction selectivity is usually achieved by adjusting the pressure of the supercritical fluid at constant temperature or, less often, by changing the temperature of the supercritical fluid at constant pressure.

The extraction selectivity of the supercritical fluid used may also be changed by the addition of small amounts of modifier, such as water or methanol (usually under 2%), directly to the sample matrix as an easier alternative to adding the modifier at a constant level to the supercritical fluid itself. For supercritical carbon dioxide the addition of small amounts of polar modifier to the sample matrix will increase the selectivity towards polar solutes. Selectivity towards the target solutes may therefore be enhanced by the use of both a specific pressure/temperature regime and the addition of polar modifier to the sample matrix or mobile phase.

In addition to the above, major advantages of SFE directly coupled to packed column SFC are:

(i) Its suitability for trace analysis — quantitative transfer of solutes between extraction and chromatography with little or no loss of (particularly) volatile solutes compared to off-line SFE-SFC or other instrumental analysis.
(ii) Ease of selective sample preparation which is easy to automate with the minimum of operator involvement.
(iii) Less contamination potential for trace level solutes, as a result of the direct coupling of extractor and chromatography.
(iv) The ability to rapidly extract and directly analyse unstable and oxidation sensitive solutes such as metal carbonyl compounds [27], which decompose on contact with most solvents, and tocopherols from wheat germ samples [28] sensitive to oxidation. For metal carbonyls and tocopherols off-line SFE-SFC would be difficult, if not impossible.

The direct coupling of SFE to SFC is nearly always achieved through the on-line linkage of a SFE vessel through a valve switching system, although other means for the direct introduction of SFE extracts such as the thermal modulation interface, developed by Mitra and Wilson [29], have been used for capillary SFC. The thermal modulator, which is interfaced between the supercritical extraction cell and SFC, is made of a short length of fused silica capillary tubing containing stationary phase on the inside and painted on the outside with an electrically conducting paint. As the supercritical extraction media flows through the modulator, the solutes of interest are trapped and concentrated in the stationary phase. The solutes are then rapidly released, like an injection, to the SFC column when the modulator is rapidly heated by passage of electrical current through the paint.

Integrated instrumentation for SFE coupled with SFC is available commercially through such manufacturers as the Suprex Corporation, Japan Spectroscopic Co. Ltd. (JASCO), Carlo Erba Strumentazione and Dionex. Commercial SFE extraction systems which include high pressure pumps, extraction cells, back pressure regulators and valves for off-line sample collection, together with the necessary heating units to maintain the supercritical fluids at temperatures above critical, are also increasingly available through a number of manufacturers including the above and ISCO, Gilson and Hewlett-Packard. However, many SFE-SFC systems used today have been constructed by workers using component parts from the above manufacturers together with commercially available sample switching valves and extraction vessels. The extraction vessels may be adapted from HPLC columns or specially prepared and developed for SFE for a particular sample matrix.

A major advantage of using packed column SFC in conjunction with SFE is the sample loading capacity of the columns. This in turn enables enough material to be chromatographed to make fraction collection of the components of interest a viable option and so fraction collection is often added to, or an option for, SFE directly coupled to packed column SFC.

A typical configuration for SFE coupled to packed column SFC is shown in Figures 5.3 and 5.4. These were chosen to illustrate typical configurations for an SFE-SFC system in terms of the supercritical flow path. The figures show

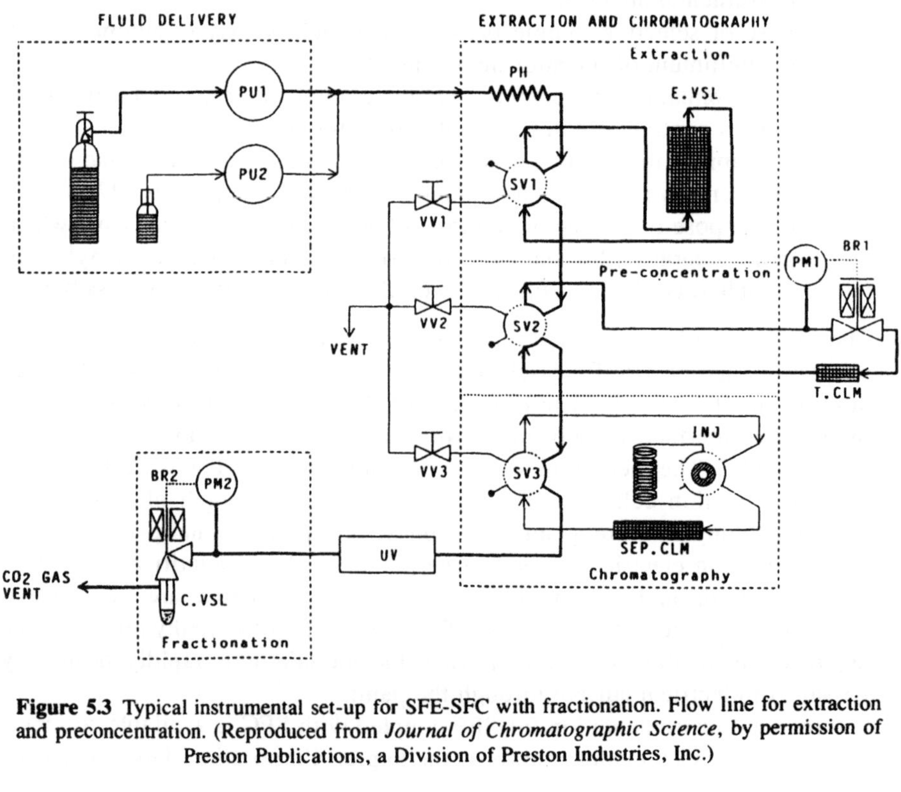

Figure 5.3 Typical instrumental set-up for SFE-SFC with fractionation. Flow line for extraction and preconcentration. (Reproduced from *Journal of Chromatographic Science*, by permission of Preston Publications, a Division of Preston Industries, Inc.)

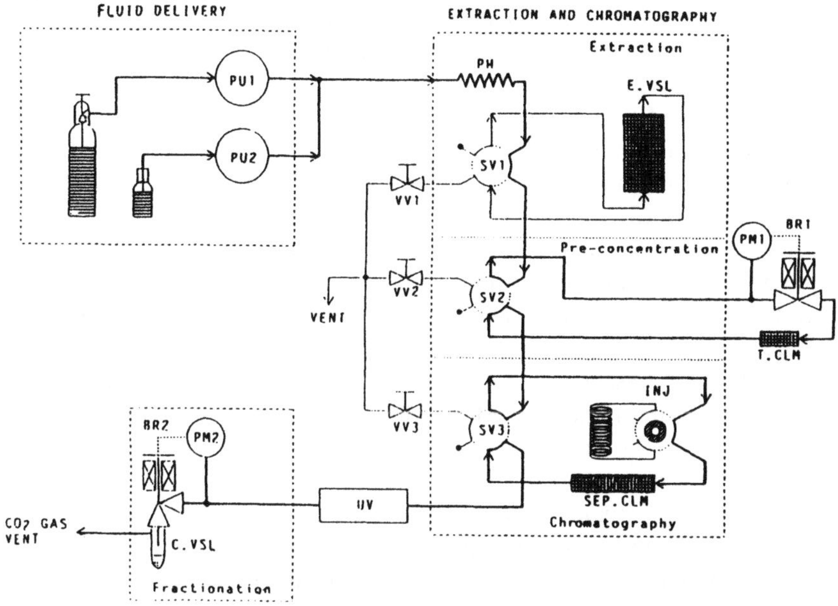

Figure 5.4 Typical instrumental set-up for SFE-SFC with fractionation. Flow line for SFC and fractionation. (Reproduced from *Journal of Chromatographic Science*, by permission of Preston Publications, a Division of Preston Industries, Inc.)

the position of post-SFE trapping (or pre-concentration as shown in the figures) and post-SFC fractionation. Figure 5.3 illustrates the mobile phase flow path during the extraction and pre-concentration stage, while Figure 5.4 shows the flow path during SFC (and fractionation). In the figures given, UV is used for solute detection. As discussed previously, UV detection is commonly used for packed column SFC, however IR and FTIR are also frequently used and gas detectors such as the flame ionization detector (FID) can be used when an eluent splitter is introduced post-column. In Figures 5.3 and 5.4 the dotted boxes indicate the three different sections of the SFE-SFC system: fluid delivery (carbon dioxide and modifier cylinders plus fluid delivery pumps); extraction, including pre-concentration and chromatography; and fractionation.

During the extraction process liquefied carbon dioxide, indicated by the thick flow line in Figure 5.3, is pumped via the high pressure pump (PU1) through the preheat coil (PH) where the carbon dioxide is heated to a temperature above critical. The pressure of the pumped carbon dioxide then builds up to the pressure to be applied to the extraction vessel (E.VSL), above the critical pressure where it is controlled by the back pressure regulator (BR1), while the pressure of the trap column (T.CLM), placed downstream of the regulator, is kept at a lower pressure by the second back pressure regulator (BR2). On passing from the extraction vessel to the trap column the density of the now supercritical fluid carbon dioxide is decreased rapidly, which results in a decrease of its solvating power with consequent deposition of solutes onto the packed bed of the trap column.

On completion of extraction and trapping of the solutes, the extraction vessel is bypassed by means of the switching value (SV1) and supercritical fluid, which can also contain a modifier added via pump (PU2), is passed through the trap column to flush the solutes in a concentrated band to the separator column (SEP.CLM) via the first back pressure valve which is now completely open. The flow path for SFC is indicated by the thick flow line in Figure 5.4. During chromatography, the pressure of the carbon dioxide in the entire flow line from the pumps through to the detector is controlled by back pressure regulator (BR2).

After SFC the pressure of the column eluent is reduced to atmospheric on passing through the regulator (BR2) which drastically reduces the solubility of the solutes in the carbon dioxide, enabling them to be collected by an appropriate collection device, such as the one shown.

In SFE-SFC the separation and trapping are housed within an oven so that temperatures above the critical temperature may be maintained. The SFE vessel may be housed in the same oven used for the separator column but for versatility the extraction vessel is usually kept in a separate oven with an independent temperature control. Alternatively, the extraction cell may be thermostated in some other way to maintain a constant temperature during extraction.

Selectivity of a supercritical fluid towards target solutes may be at least partially achieved through variation of its solvating power, through the control of pressure (density) in the system. The higher the density of supercritical fluid carbon dioxide, the greater its solvating power. A change in solvating power may also be achieved by variation of extraction temperature but this is less commonly used.

Coupled SFE-SFC is often configured with a solute concentration device prior to valve switching onto the SFC. The objective of this device is to concentrate the solutes of interest into a small band for 'injection' onto the SFC. There are two main approaches to this concentration stage. The first is to use an 'extractor' column that is packed with absorbent material, such as silica, on which the solutes are trapped. Saito *et al.* [28] used this type of column to collect extracted tocopherols from wheat germ. The second approach is the use of cryofocusing. This technique has been applied by a number of groups, in particular Ashraf-Khorassani *et al.*, who used the technique to trap polymer additives prior to valve switching onto SFC [30–32], Andersen *et al.*, who used it to extract cold-pressed grapefruit oil flavorants [33], and Xie *et al.* [34], who used SFE with cryogenic trapping for the extraction and collection of a standard hydrocarbon (octatriacontane) and the biologically active compound oubain in conjunction with packed column SFC and a fraction collector.

5.4 Instrumental aspects

5.4.1 Back pressure regulators

In SFE and SFC back pressure controllers are required to maintain the pressure of the mobile phase above its critical point. Two types are used: the simple capillary restrictor or the mechanical regulator which may be electronically controlled.

The simple restrictor is most commonly used for capillary SFC, generally in conjunction with gas detectors or for small volume extraction vessels in SFE. It consists of a short length of narrow bore fused silica capillary tubing. Pressure regulation is achieved through flow resistance brought about by the length and the diameter of the capillary tube. The longer and narrower the tube, the greater the pressure maintained. Several types of capillary restrictor have been used having linear, crimped, tapered, converging or diverging, integral and fritted ends [35]. The major advantage of these simple restrictors is cheapness, however restrictor to restrictor back pressure reproducibility is difficult and the restrictors are prone to blockage often due to condensation of moisture from the sample which becomes frozen at the tip of the restrictor on evaporation of the carbon dioxide eluent. Capillary restrictors are useful for the direct on-line coupling of SFE to capillary SFC or gas chromatography, in which the restrictor can be used as an on-column injection device through the injection port and directly into the separator column [36 and chapter 3].

Mechanical restrictors are better suited to the higher mass flow rate of packed column systems and for use with UV detectors. While capillary restrictors for capillary SFC are situated before gas detection, mechanical restrictors are usually situated behind the UV detection cell to maintain pressure throughout the whole SFC (or SFE-SFC) system. Mechanical restrictors consist of mechanically adjustable valves or pneumatic systems which can be electronically controlled for density programming. Electrical feedback regulators are commonly used which consist of a pressure sensing device and a needle valve. These regulators are able to control back pressure irrespective of the mass flow rate of the supercritical fluid and are consequently more reliable for both SFE and SFC where extraction or chromatographic reproducibility is a function of pressure. As a result of the high dead volumes obtained with conventional mechanical back pressure regulators and the need to use low dead volume high-precision back pressure regulators for fractionation of SFC eluted peaks, Saito *et al.* [25] developed a new back pressure regulator valve based on a different operational principle than conventional regulators (Figure 5.5). The regulator consists of a needle driven by a solanoid and a valve seat. Generally this type of valve controls the flow resistance of the fluid by changing the gap between the valve needle and seat. The new valve operates by high speed switching of the fluid flow by opening and closing the flow path. The regulator prevents blockages in the flow path which are possible

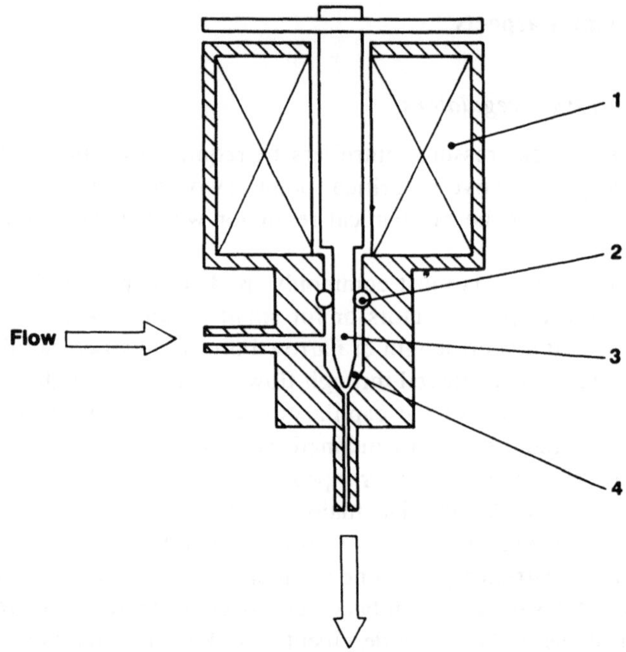

Figure 5.5 Cross-sectional view of flow switching back pressure regulator valve: 1. needle drive solenoid; 2. needle seal; 3. valve needle; 4. valve seat. (Redrawn with permission from the Royal Society of Chemistry [25].)

because precipitated solutes and dry ice from the carbon dioxide are continually trapped and forced to pass through the valve. To maintain the required back pressure the valve is used with a pressure transducer and control circuitry. Trials showed that the regulator was able to maintain pressure settings of supercritical carbon dioxide to within 1% between pressures of 100 and 350 kg cm^{-2}.

5.4.2 Extraction vessels

There are a number of commercial supercritical extraction systems, marketed by Jasco, Suprex, Dionex, ISCO, Hewlett-Packard and Carlo Erba, which may be purchased either as an integrated instrument with SFC or separately and linked by the user to SFC.

Commercial SFE instrumentation employs cells ranging from a few microlitres to tens of millilitres. However, many supercritical fluid extractions are performed in pressure rated HPLC columns with perhaps the most common size being the standard 250 × 4.6 mm i.d. analytical HPLC column.

HPLC columns as extraction vessels for SFE are readily obtainable and can be packed and unpacked so that they can be reused. McNally and Wheeler [37] have used a variety of column sizes as extraction columns for SFE-SFC:

commonly 250 × 4.6 mm i.d. and 100 × 4.6 mm i.d. columns. They also used 250 × 1.27 mm i.d. hollow stainless steel tubes and a stainless steel bolted closure pressure vessel to extract soil, plant materials and cell culture medium. The sizes of the vessels used depends on the quantity of material to be extracted, detection limits required and the back pressures obtained during extraction. Vessels having volumes over 3 ml were pressure-rated to withstand pressures up to 4200 psi at 100°C. Standard HPLC columns have been used by many other workers including Xie *et al.* [34] and Ramsey *et al.* [38]. In their paper on supercritical fluid extraction as a means of sample introduction for chromatography, Andersen *et al.* [33] discuss volume considerations for SFE extraction vessels, particularly in terms of requirements for environmental analysis. In their laboratory they used extraction vessels with volumes ranging from 0.3 to 5 ml but they felt there was a requirement to build extraction vessels capable of extracting the 50 g of material specified in most current methods for environmental analysis. Vessels having an extraction capability for 50 g of leaf sample have been built and are referred to in the paper.

During extraction the sample, usually ground for solid material, is packed into the extraction vessel and restrained by frits. Where liquids or oils are to be extracted they may be deposited on a packing material such as silica, quartz wool or filter paper. The latter was used by Nishikawa [39] to examine extraction parameters for diniconazole pesticide standards, using filter paper spiked with the pesticide prior to SFE-SFC.

Micro and mini extraction vessels having void volumes of 3–4 μl and 85 μl respectively were built by Jahn and Wenclawiak [27] for direct on-line coupling of SFE with packed column SFC for use under sub- and supercritical conditions. The advantages of using such small extraction volumes were seen as rapid equilibrium between the analyte samples and the supercritical extraction fluid, and minimal disturbance of the column packing material on direct transfer to the analytical column. The micro and mini extractors used are illustrated in Figures 5.6 and 5.7.

The micro extractor was constructed from a Rheodyne in-line solvent filter. Short connections were made from 0.1 mm i.d. tubing and 1/16″ fittings. The 5 μm frit used in the micro extractor was prepared by compression with a conical tool to create a small depression for the sample material. A filter was

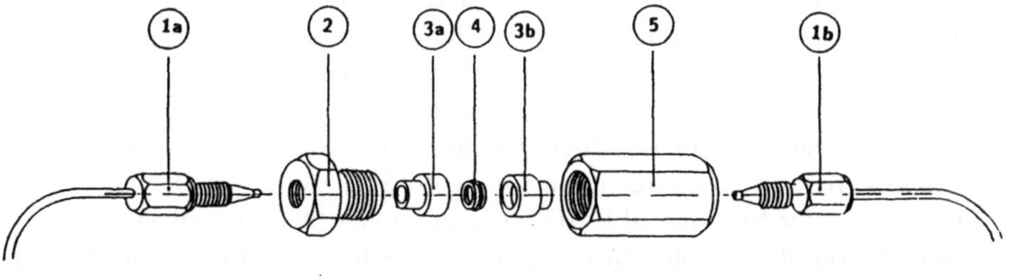

Figure 5.6 Micro extractor: 1a,b. 1/16″ male nuts with ferrules; 2,5. casing; 3a,b. female connection and frit holder; 4. frit. (Reproduced with permission from *Chromatographia* [27].)

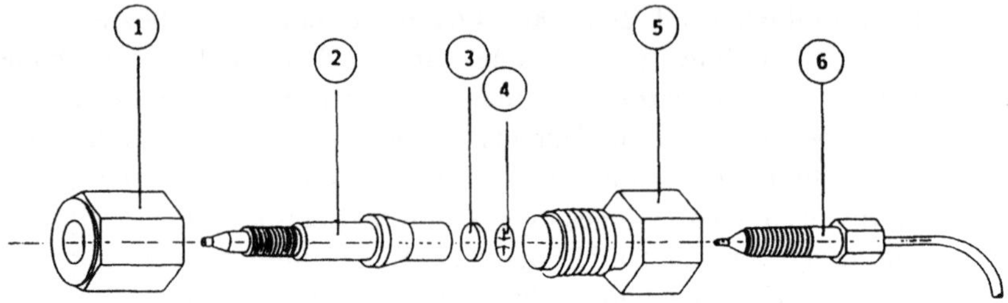

Figure 5.7 Mini extractor: 1. nut; 2. extraction cell; 3. frit; 4. stream distributor; 5. bolt; 6. extra long 1/16″ bushing. (Reproduced with permission from *Chromatographia* [27].)

placed into the frit holder to prevent access of sample material to the column during SFE. PTFE and cellulose acetate (pore sizes 0.2 and 0.45 μm) were found suitable as filter materials since they could withstand supercritical carbon dioxide at temperatures up to 80°C. The micro extractor was resistively heated via the capillary extractors (0.4 V, 180 W adjustable) and used to extract highly viscous and solid samples, such as used engine oil, for benzene, toluene and naphthalene, and solid Cr, Mo and W mesitylene carbonyl compounds which react with normal solvents and are also thermally unstable.

The mini extraction vessel, shown in Figure 5.7 was constructed from a guard column having an internal diameter of 2.1 mm, a length of 25 mm and an internal volume of approximately 85 μl. The mini extraction vessel was mainly used for supercritical extraction from solid materials such as soil, soot or from liquids deposited onto adsorbent materials, such as C18 silica. The mini extractor was fitted with a filter plate on the inner side of the frit. After filling the extraction vessel with the material to be extracted it was screwed directly to the sampling valve via capillary tubing. The micro extractor was heated by the attachment of an aluminium clamp containing a 70 W GC detector heater.

As seen from the above, there is little difficulty in extracting solid samples by SFE. Viscous samples, such as used engine oil, may also be introduced directly for SFE-SFC provided the sample is very small.

Liquid samples present more of a problem. The usual solution is to place the liquid in an extraction vessel by injection onto solid adsorbent material, either in small microlitre volumes or larger volumes if a volatile organic solvent is used and allowed to evaporate before extraction.

Aqueous samples (and samples containing a high proportion of water) present special problems due to the difficulties of water freezing in the SFE-SFC flow path creating potential blockages, and the effect of water, acting as a polar modifier, on solute retention time with SFC using chemically bonded silicas. Thus for SFE and SFE-SFC it is important that most of the water is removed from the sample. This may be done before SFE by evaporation, by use of a solid adsorbent such as anhydrous sodium sulphate or by freeze drying, which is commonly used for biological tissues [38].

SFE of aqueous samples has an important potential for aqueous environmental samples, for the analysis of pesticides and herbicides from field drainage and municipal waste water, and for the pharmaceutical industry for the analysis of water soluble drug formulations and metabolites [40] where the direct introduction of aqueous samples for SFE or SFE-SFC would be very advantageous. Two approaches to the direct introduction of aqueous samples for SFE-packed column SFC are: (i) to use an on-line aqueous extraction vessel, as reported by Hedrick and Taylor [40], or (ii) to use a phase-separator, to remove water post-SFE pre-SFC, as described by Thiebault *et al.* [41].

Hedrick and Taylor extracted a phosphonate (diisopropyl methylphosphonate-DIMP) from aqueous media using the extraction vessel obtained from the Suprex Corporation (Pittsburgh, PA, USA) as shown in Figure 5.8. 3 ml of aqueous sample is introduced into the 10 × 1 cm i.d. extraction vessel which has an internal volume of 8 ml. With the extraction vessel located in an oven, supercritical carbon dioxide is passed into the bottom of the vessel where it bubbles through the aqueous media to extract the compound or compounds of interest. As a result of the lower density of supercritical carbon dioxide compared to water, the supercritical fluid rises to the top of the extraction vessel

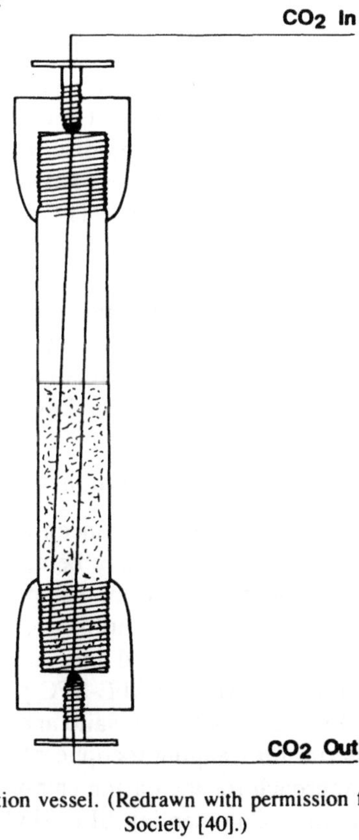

Figure 5.8 Aqueous extraction vessel. (Redrawn with permission from the American Chemical Society [40].)

where it is recirculated through the aqueous media for the determined extraction time. A 20 μl extraction aliquot is then transferred to the SFC column via a sample loop for quantitation. With this method of on-line extraction for aqueous samples Taylor and Hedrick were able to obtain a linear four point calibration curve over the concentration range of 834 ppb to 834 ppm for DIMP, using FTIR detection. The rate of extraction of DIMP from water was determined using SFC/FID response: an increase in the amount of material extracted with time was found up to equilibrium at 1.5 hours. However, addition of 0.1 mg of sodium chloride to the aqueous extraction medium reduced equilibrium time to less than 5 minutes.

Use of the aqueous extraction vessel was also extended to the qualitative study of various hydrochloride salts of drug compounds including triprolidine hydrochloride by SFE-SFC.

In the second approach to the extraction of solutes from aqueous media Thiebaut *et al.* [41] used a phase-separator to demonstrate the potential of segmented-flow systems for on-line liquid-SFE and SFE-SFC as shown in Figure 5.9. This figure illustrates the location of the separator in the flow path for coupled SFE-SFC. Interior details of the phase-separator are shown at the bottom of the figure. The aqueous extract together with supercritical carbon dioxide passes along the grove in the separator, which has a upper hydrophobic and a lower hydrophilic (usually stainless steel) surface where separation of water and supercritical carbon dioxide occurs based on the respective 'wettability' of the hydrophilic stainless steel part and hydrophobic part of the phase-separator's construction. For SFE, the disc material functioning as the hydrophobic surface in the separation process has to be carefully selected to withstand the pressure conditions of SFE. PTFE was found unsuitable but good stability was achieved with Delrin and PVDF which did not show any signs of deterioration after several months.

Initially the system was used to extract phenol and 4-chlorophenol from water and phenol from urine. Extraction efficiency of the test compounds was over 85% with a repeatability of about 8% relative standard deviation for 8 samples for the total SFE-phase separation-SFC system.

5.4.3 On-line solute trapping and concentration

After SFE there is often a need to further concentrate the solutes of interest before sample introduction to SFC. This may be achieved by repeated evaporation of supercritical carbon dioxide in the sample loop prior to injection, so that a greater concentration of material builds up, or by trapping the solutes onto a suitable adsorbent material packed into a sample loop or small or accumulator column in the flow path for SFE-SFC prior to SFC. This method has been used by several workers including Saito and Yamauchi [28] who used silica as a trapping material for trapping tocopherols.

A third and popular approach for increasing sample concentration is to use on-line cryogenic trapping. Commercial SFE-SFC systems incorporating

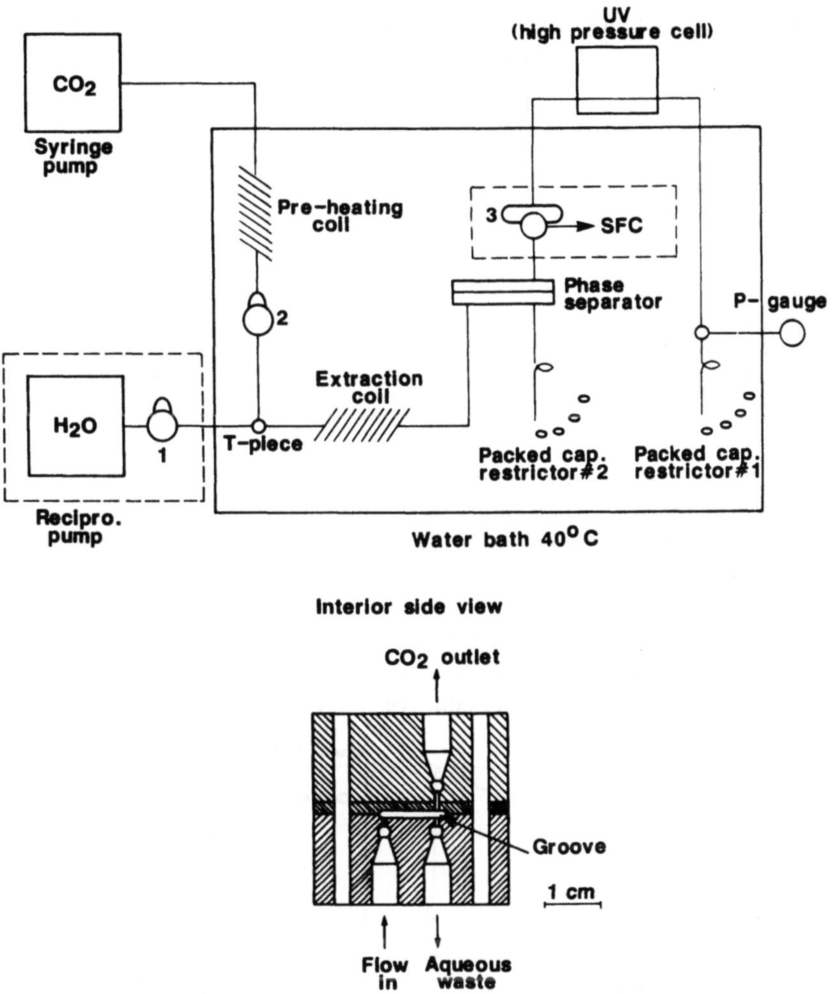

Figure 5.9 Schematic of SFE-phase separation-SFC to show the position of the phase-separator. Insert: expanded interior view of the phase separator. (Redrawn with permission from *Journal of Chromatography* [41].)

cryogenic trapping facilities are available from Lee Scientific/Dionex, Suprex, Jasco and others. The main advantage of using cryogenic trapping for coupled SFE-SFC is the ability to concentrate the analytes of interest into a discrete narrow band, before injection via a sampling valve, to give quantitative and reproducible results with better detection limits. Cryofocusing also reduces possible interaction between solutes and adsorbent when the latter concentration system is used.

Cryogenic trapping or focusing has been used by a number of workers including Ashraf-Khorassani and Levy [30] who demonstrated the potential of

SFE-cryogenic trapping-SFC for the quantitative analysis of polymer additives in low density polyethylene using a commercial Suprex system consisting of an extraction vessel, two electronically actuated valves, a zero volume T-piece, a cryogenic trap and a packed SFC column. The stainless steel cryogenic trap measured 30 × 1.0 mm i.d. and had the potential for rapid cooling and ballistic heating.

On extraction the SFE extract is passed through a 50 μm i.d. fused restrictor and into the cryogenic trap which can be cooled to as low as – 50°C using industrial carbon dioxide to collect the solutes. During extraction depressurized carbon dioxide mobile phase is vented through the ten-port valve in the chromatography/extraction oven.

Following extraction the pressure automatically changes from that set for extraction to that required for commencement of SFC. During this period the extracted solutes remain in the cryogenic collection trap until pressure equilibrium for chromatography is reached when the ten-and four-port selector valves automatically switch to the injection mode. The temperature of the cryogenic collection trap is then rapidly heated to the injection temperature at a rate of 250°/min. The supercritical carbon dioxide from the pump then passes through the cryogenic trap when the extracted and concentrated components are transferred to the SFC column for quantitation. Using the system described, Ashraf-Khorassini and Levy were able to obtain linear calibration curves for four different polymer additive standards added to polymer pellets,

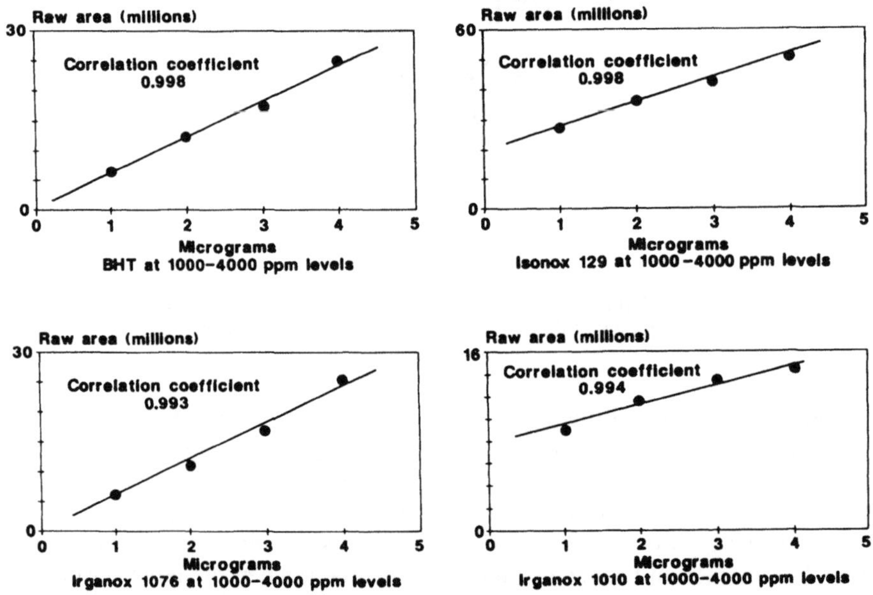

Figure 5.10 SFE-SFC calibration curves for BHT, Isonox 129, Irganox 1076 and Irganox 1010 at various spike levels. (Redrawn with permission from the *Journal of High Resolution Chromatography* [30].)

BHT, Isonox 129, Irganox 1076 and Irganox 1010 at the concentrations given in Figure 5.10.

In a second paper Ashraf-Khorassani *et al.* [31] evaluated the SFC-cryogenic trapping-SFC system with two different hydrocarbon standard mixtures, while in a third [32] the effects of temperature, matrix configuration and extraction time were studied.

Andersen *et al.* [33] discussed aspects of the use of cryogenic traps, particularly with respect to trapping temperature for volatile solutes and some of the problems involved. It was generally observed that trapping becomes more difficult with increasing solute volatility, showing that the efficiency with which the solute is trapped is a function of temperature. Cryofocusing of extracts containing significant water content requires special precautions to prevent ice crystals accumulating in the trap. Aspects of cryofocusing were demonstrated with essential oils.

Xie *et al.* [34] constructed a SFE capillary column SFC system with cryogenic trapping (and fraction collection) to evaluate its use with a biologically active compound and a standard hydrocarbon, normal octatriacontane. These workers also found that, in general, the lower the cryofocusing temperature the better the result. By altering trapping temperature selective trapping of solutes was possible.

5.4.4 On-line sample introduction

For SFE directly coupled to packed column SFC, introduction of SFE extracted solutes to the packed SFC column after they have been trapped in a sampling loop, concentrated onto an adsorbent (accumulator) column or trapped cryogenically is invariably achieved by the use of a high pressure rotary sample introduction valve, having either manual or, far more commonly, electronic or pneumatic actuation. In the case of cryogenic trapping the trap may also be heated rapidly to ensure the rapid transfer of a narrow band of solutes to the head of the SFC column through the open sampling valve. A further possibility for sample transfer between extractor and SFC is to use thermal modulation [29] but this has only been used to date for the transfer of solutes to capillary and not packed column SFC.

Direct transfer of solutes from SFE to SFC using supercritical carbon dioxide is a major attraction of using SFE directly coupled to SFC. Off-line SFE usually involves collection of the solutes in an organic solvent, which must be reintroduced into the supercritical fluid for SFC. This often results in the incomplete mixing of solvent/supercritical fluid leading to band broadening and loss of chromatographic efficiency. In addition, traces of the organic solvent containing the solutes may be left in the SFC sampling valve, resulting in memory effects. Direct coupling of SFE with SFC avoids these incompatibility problems and results in greater chromatographic efficiency, especially for trace analysis.

As many commercial sampling valves are available which may be used under the pressure and temperature conditions for SFE-SFC analysis, they will not be discussed further here, however it is important in SFE-SFC, as with all other forms of chromatography, to ensure that dead volumes are minimised or as far as possible eliminated, if good reliable chromatographic results are to be obtained.

5.4.5 Columns

Materials for packed columns used for SFC have been discussed earlier. For SFC, many workers still use conventional HPLC packing materials based on silica which, as previously discussed, may give some problems with polar solutes due to the surface activity of these materials. To overcome this problem, increasingly inert column packing materials for SFC are being developed and used. Polymer packing materials and porous graphitic carbon columns which have little or no surface activity show increasing promise for packed column SFC, although at present there may be some difficulties with their use at the elevated pressures used for SFC.

The availability of a wide range of selective HPLC packing materials, including materials for enantiotropic separations, together with new packing materials being developed for SFC is a major advantage of packed over capillary SFC where there is a smaller range of surface coating materials.

Virtually all standard analytical HPLC column sizes and types can be used for SFC applications, provided that the columns have been sufficiently pressure rated to withstand the intended SFC operating pressures. Where increased resolution is required longer packed capillary columns may be used. When sample loading is an important consideration larger or semi-preparative columns may be used, particularly when fraction collection of eluted solutes is required.

5.4.6 Detectors

Detectors used for SFE coupled with packed column SFC have been discussed earlier. In general, packed column SFC lends itself to the use of liquid (HPLC) type detectors, particularly the UV and diode-array or multiwavelength UV detectors which are used by most workers. UV detectors have the advantage of being compatible with the use of polar modifiers, such as methanol which causes a high background signal with FID, added to the supercritical carbon dioxide mobile phase. Infrared or FTIR detectors are also fairly commonly used for packed column SFC. Gas (GC) type detectors may also be used, after a column effluent splitter to reduce the flow, in particular FID which is also fairly commonly used for packed column SFC, although other GC-type detectors may be used.

Mass spectrometry has also been used for detection of eluted solutes from packed column SFC by Ramsey *et al.* [38] where MS-MS was used to detect veterinary drug residues extracted by SFE from freeze dried pigs kidney.

The major difference between the liquid and gas type detectors used for SFC is that supercritical conditions for the mobile phase are maintained for the former through the detector, with the restrictor placed behind the detector, while for the latter the restrictor is placed before the detector which works under ambient pressure. Because of this difference it is generally necessary to modify liquid-type detectors to withstand the higher pressures used for SFC.

5.4.7 *Fraction collection*

A major consideration in using SFE combined with packed column SFC is the greater sample capacity (discussed earlier) of packed vs. capillary columns, particularly when fraction collection of eluted solutes is required for further investigation. Saito and Yamauchi have described a number of fraction collection devices for SFC in their handbook [42] and described a system for fraction collection in their paper on enrichment of tocopherols in wheat germ [28]. The thick line in Figure 5.4, taken from this paper, shows the flow path through an SFE-SFC system for fraction collection. The extraction vessel is bypassed by means of the switching valve (SV1) and supercritical carbon dioxide, in this case containing modifier, flows through the trap and separation column (SEP.CLM) and through the UV detector via the first back pressure regulator (BR1), which is now open. The second back pressure regulator (BR2) which is used to maintain the pressure throughout the flow path, including the trap and separation columns and UV detector, is used to release the pressure of the column effluent to atmospheric, which causes a sharp decrease in solute solubility, enabling the collection of individual solutes into glass collection vessels, which in this case were changed manually.

McNally and Wheeler [37] investigated a number of collection devices from SFE packed column SFC in their determination and fractionation of sulfonyl urea herbicides and their metabolites. During initial SFE-SFC experiments, radiolabelled metabolites were extracted from various matrices, collected and detected in a liquid scintillation counter. As radiolabelled compounds were used it was necessary to achieve quantitative recovery of eluted species and a number of collection devices were investigated including two HPLC guard columns in series containing C18 packing material, a hollow stainless-steel U-tube placed in ice or a dry-ice acetone bath and a U-tube containing copper shot or stainless-steel balls. The C18 guard columns proved ineffective and less than 25% radioactive recovery was achieved from the U-tube devices. McNally and Wheeler's solution was to use a simple glass collection vessel with which collection efficiencies of up to 100% were obtained. Supercritical effluent from the separator column was passed through a final back pressure regulator, where the pressure was reduced to decrease the solubility of the solutes, and through a (solvent reservoir) frit into a glass reservoir where the radioactive materials were collected. The reservoir was fitted with a side port, to ensure pressure release during operation, which was plugged with glass wool as a precaution against radioactivity loss.

5.5 Applications

SFE coupled with packed column SFC has found many applications in the last seven years, particularly in connection with the determination of food compounds, drug residues, herbicides and pesticide residues and polymer additives. Some of these have been mentioned in previous sections of this chapter to illustrate specific aspects or uses of the technique. These, plus additional applications, will now be discussed.

Sugiyama *et al.* [10] initially demonstrated the technique of SFE coupled to packed column SFC by extracting and determining caffeine from green coffee beans with supercritical carbon dioxide. After extraction of the powdered beans (with 20% added water) at 200 bar and 40°C for 15 minutes, extracts were chromatographed on a C18 HPLC column. Caffeine was detected using a multiwavelength UV detector.

Tocopherols in wheat germ were also investigated by Saito and Yamauchi [28] using the SFE-SFC system illustrated in Figures 5.3 and 5.4, in which tocopherols were extracted from 3 g of wheat germ containing approximately 0.03% tocopherols and 10% oil by weight. Extraction was performed at 250 bar and 40°C. Separation of tocopherols was achieved using a 200 × 20 mm i.d. column packed with silica gel with supercritical carbon dioxide plus ethanol modifier as mobile phase. The eluent was monitored with a multiwavelength UV detector and the tocopherol content in each fraction collected was determined.

Engelhardt and Gross [43] used SFE-SFC to extract and determine compounds from food products and spices, such as from curry leaves, and carvone and terpene from caraway seeds. A modified flame ionization detector was used where part of the column effluent fluid is expanded through a small orifice directly into the flame. The short expansion path from column to flame prevented solute condensation and consequent detector spiking. 1–2 g of finely ground material was dry packed into the extraction vessel and extracted with supercritical carbon dioxide. After extraction the solutes were passed via a sample loop/valve to the packed SFC column which was filled with either silica or C18 packing material. Using the system described, completeness of extraction and extraction kinetics of carvone and terpene from caraway seeds were studied.

The applicability of SFE packed column SFC to the determination of lipid peroxide levels in foods and fats was investigated by Sugiyama *et al.* [44]. Good agreement was found between results from SFE-SFC and potentiometry previously used for hydroperoxide compounds extracted from samples of ground peanuts. SFE-SFC also had the advantage of providing an oxygen free environment for the analysis of these sensitive thermally labile compounds.

Fatty acids, fatty acid esters and tocopherols were extracted and analysed by Ikushima *et al.* [45] using SFE-SFC and FTIR detection. After extraction the solutes were separated on a packed column containing silver nitrate on silica. A heated transfer line was used to transfer the separated compounds to the FTIR detector using Ga-Gd-garnet windows specially selected for the work at supercritical carbon dioxide pressures. The separation and efficiency determined for this system were correlated with a parameter derived from the solubility parameter concept.

Veterinary drug residues including trimethoprim, hexestrol, diethylstilbestrol and dienestrol, were determined by SFE-SFC-MS-MS from pigs kidney by Ramsey *et al.* [38]. As a result of the quantity of endogenous material obtained from powdered freeze dried kidney, SFE-MS or SFE-MS-MS was considered impractical due to the likelihood of rapid contamination of the mass spectrometer. To avoid this difficulty, a SFE-SFC procedure was developed to provide additional clean-up before MS-MS. SFC was performed on a polar amino bonded column; initially using 100% supercritical carbon dioxide followed by sharp methanol gradient to 80:20 CO_2/MeOH in 0.5 minutes. The final composition was then maintained for 10 minutes. The methanolic (modifier) gradient was developed to exploit the difference in polarity between polar drug residues and the generally non-polar endogenous material co-extracted during SFE to obtain maximum chromatographic separation. Although extraction efficiencies for the investigated drugs were not established, re-analysis of the dried extraction residue showed no further extractable drugs. The authors concluded that SFE-SFC-MS-MS had considerable potential for the determination of trace levels of contaminants in foods and other matrices.

The potential of SFE packed column SFC for the determination of prostaglandins in hydroxypropyl methylcellulose (HPMC) dispersions, widely used for controlled-release drug formulations, was demonstrated by Roston [46]. Quality control procedures for formulated drug substances usually involve lengthy clean-up procedures to remove the drug of interest from the formulation matrix. Using SFE-SFC the only sample preparation necessary was to weigh the material into the extraction vessel. After extraction the extracts were cryofocused and submitted to SFC using a cyano column. Extraction efficiency for 4 minutes for misoprostol was estimated at 65%. Total analysis time for SFE-SFC was 20 minutes. With optimization of experimental parameters it was felt that extraction efficiency could be improved and that SFE-SFC had potential for quality control procedures.

The determination of phenol and 4-chlorophenol in aqueous samples and phenol in urine was investigated by Thiebaut *et al.* [41] using a segmented flow system and phase separator after SFE and prior to SFE-SFC.

A number of environmental samples have been determined by SFE-SFC. Phosphonate was extracted and analysed from aqueous media by Hedrick and Taylor [40] using a specially adapted extraction vessel. McNally and Wheeler [37] used SFE coupled with packed column SFC to determine sulfonylurea herbicides and metabolites from complex matrices including soil, plant materials and cell culture medium. Near-quantitative or quantitative recovery was achieved for radiolabelled metabolites eluted from SFC and collected in a special glass collection vessel. SFE-SFC has been used for the direct extraction, clean-up and determination of pesticide residues by Warner and Yancey [47]. SFE-SFC, using supercritical carbon dioxide, had advantages over conventional methods of pesticide residue analysis principally because supercritical carbon dioxide is an efficient solvent for the non- or moderately polar pesticide residues and a poor solvent for the highly polar carbohydrate and proteinaceous materials which are co-extracted and interfere with conventional GC-MS pesticide residue analysis from non-oily agricultural crops. SFE-SFC was seen as being amenable for automation and many non-volatile and thermally labile pesticides could be determined without derivatisation by selective ionization detectors and MS available for SFC.

Nishikawa [39] also applied SFE-SFC to pesticide residue analysis from plant samples and soils. Three representative categories of pesticides were investigated: an organophosphorus pesticide, a synthetic pyrethroid and an azole compound. Extraction conditions, pressure, temperature and time were optimized by spiking standard compounds onto filter paper prior to SFE. Although good reproducibility was obtained for the compounds from filter paper initial results from soil were variable. Investigation showed that results were dependent on sample homogeneity and preparation. Extraction of fenitrothion spiked on powdered and from tableted soil gave relatively low RSD values but the tablet method improved extraction efficiency as a result of the homogenous fixing of the pesticide. However, extraction efficiencies were lower than those obtained using filter paper. It was observed that the pesticides investigated gave individual extraction efficiencies which were dependent upon their electronic structures. The SFE-SFC of pesticide residues from straw, green tea and vegetables was also investigated.

A variety of polymer additives has been quantitatively investigated by Ashraf-Khorassani *et al.* [30–32] using the technique of SFE-cryogenic trapping-SFC. Extraction efficiencies of polymer additives from a number of different matrices were examined by varying extraction conditions. Polymer additives from low density polyethylene (LDPE) as well as propellant stabilisers and normal hydrocarbons have been examined by the hyphenated technique.

Polyethylene and polypropylene samples from several manufacturers were extracted by SFE coupled to SFC by Ryan *et al.* [48]. Extraction efficiencies of greater than 92% were obtained for ten different additives including butylhydroxytoluene (BHT) and Irganox 1010. SFE-SFC had the advantage over conventional determinations for polymer analysis of speed and efficiency.

Off-line liquid extraction of polymer additives involved many time-consuming steps including lengthy Soxhlet extraction, often followed by concentration, clean-up and reconstitution for instrumental analysis by GC, HPLC or SFC. Coupled SFE-SFC for most polymer additives can generally be accomplished in less than an hour using supercritical carbon dioxide. After extraction the additives were trapped on an accumulator column containing cyano trapping material. Separations were achieved using a packed octyl column. Both FID and UV detection were used. Linear calibration curves for quantitation were constructed by extracting standards spiked onto quartz wool in the extraction vessel.

Jahn and Wenclawiak [27] used their micro and mini extractors, previously discussed, for the direct SFE-SFC of pyrolyzed wood and viscous materials, such as used engine oil, polynuclear aromatic hydrocarbons and oxygen sensitive and reactive compounds such as metal carbonyls, demonstrating the usefulness of the technique for these categories of materials.

5.6 Future prospects

The future for SFE coupled to packed column SFC looks very bright. The potential for its increased use is enormous. This potential has been demonstrated by the many and varied applications discussed in this brief chapter: in the foodstuffs industry, the polymer industry, for environmental analysis of herbicides and pesticides in soil, agricultural crops and waste and drainage waters, for the pharmaceutical industry and in many other areas where qualitative and quantitative trace analysis is required for quality control, regulatory and research purposes.

The advantages of the efficiency, speed and simplicity of SFE coupled with the ease of extracted solute concentration, with an adsorbent or cryogenic trapping system prior to direct transfer to SFC, make the technique very attractive compared to many currently used off-line processes.

Because of its potential for aqueous samples, it is likely there will be intense activity in developing systems capable of extracting these samples and samples previously very difficult to extract such as biological fluids and tissue.

SFE-SFC using supercritical carbon dioxide has, because of its inertness, great advantages and potential for use with reactive and heat labile compounds such as metal carbonyls which react with most organic solvents and for substances liable to oxidation which would be difficult to extract and analyse by other means.

High selectivity towards the solutes of interest can be achieved with SFE coupled to packed column SFC as a result of the extraction selectivity provided by pressure and temperature parameters and the possible use of modifiers in the supercritical carbon dioxide, with additional selectivity provided by packed column SFC through the wide range of commercially available column packings with different selectivities.

Two further attractions of SFE coupled with packed column SFC are the sample capacity of packed columns, which makes the technique valuable for the fraction collection of solutes for additional investigation, and the very wide range of detectors that may be used. Both liquid and gas type detectors can be used and in addition may be hyphenated with mass spectrometry or MS-MS to provide a highly selective and sensitive analytical tool. Further development of detectors specially suited to SFC and SFE-SFC are likely as are more commercial SFE instruments to be used with SFC, GC or HPLC.

Further work is needed in the development of inert column packing materials. Although polymer coated columns specially developed for packed column SFC are commercially available there is a need for a greater choice of materials for SFC. Further developments from porous graphitic packing material look likely, in order to develop a totally inert packing material, as do developments with materials for polymer columns.

Finally there is a great interest in hyphenating other analytical techniques to SFE, SFC and SFE-SFC, particularly FTIR and MS and this trend is likely to continue in the analysts' quest to know ever more about less and less!

References

1. J. B. Hannay and J. Horgarth, *Proc. Royal. Soc. (London)* **29** (1897) 324.
2. K. Zosel, Australian patent Application 16/4/63.
3. T. P. Zhuze, G. N. Jushkevic and J. E. Gekker, *Maslo-Zhir. Promst.* **24** (1958) 34.
4. E. Stahl and W. Schiltz, Extraction with supercritical gases in coupling with thin layer chromatography. *Z. Anal. Chem.* **280** (1976) 99.
5. K. K. Unger and Raumeliotis, On-line high pressure extraction-high performance liquid chromatography. *J. Chromatogr.* **282** (1983) 519.
6. R. D. Smith, H. R. Udseth and B.W. Wright, *Supercritical Chromatography*, eds J. M. L. Penninger, M. Radosz, M.A. McHugh and V.J. Krukonis, Elsevier, Amsterdam (1985) pp. 191–223.
7. S. B. Hawthorne and D. L. Miller, Extraction and recovery of organic pollutants from environmental solids and tenax-GC using supercritical carbon dioxide. *J. Chromatogr. Sci.* **24** (1986) 258.
8. E. Klesper, A. H. Corwin and D. A. Turner, High pressure gas chromatography above critical temperature. *J. Org. Chem.* **27** (1962) 700.
9. K. Sugiyama, M. Saito and A. Wada, *Japanese patent.*
10. K. Sugiyama, M. Saito, T. Hondo, and M. Seda, New double-stage analysis method. Directly coupled laboratory-scale supercritical fluid extraction-supercritical fluid chromatography, monitored with a multiwavelength ultraviolet detector. *J. Chromatogr.* **332** (1985) 107.
11. K. Sugiyama, M. Saito and A. Wada, US patent application.
12. R. J. Skelton Jr., C. C. Johnson and L. T. Taylor, Sampling considerations in supercritical fluid chromatography. *Chromatographia* **21** (1986) 3.
13. M. Novotny, S. R. Springston, P. A. Peaden, J. C. Fjeldstead and M. L. Lee, Capillary supercritical fluid chromatography. *Anal. Chem.* **53** (1981) 407A–414A.
14. J. R. Wheeler and M. E. McNally, Comparison between packed and capillary supercritical fluid chromatography and high performance liquid chromatography using representative herbicides and pesticides as typical moderate polarity and molecular weight range molecules. *J. Chromatogr.* **410** (1987) 343.
15. P. J. Schoenmakers, Supercritical-fluid chromatography: open columns vs. packed columns. *J. High Resolution Chromatogr. and Chromatogr. Commun.* **11** (1988) 278.

16. M. Petersen, Review: packed columns in supercritical fluid chromatography. *J. Chromatogr.* **505** (1990) 3.
17. P. J. Schoenmakers, L. G. M. Uunk and H- G. Janssen, Comparison of stationary phases for packed-column supercritical fluid chromatography. *J. Chromatogr.* **506** (1990) 563.
18. T. Griebrokk, J. Doehl and E. Lundones, Current use of packed columns in SFC, *Progress in HPLC*, Vol. 4, eds Yoshioka *et al.* (1989) pp. 53–71.
19. L. T. Taylor and C. H-C. Chang, Packed column development in supercritical fluid chromatography. *J. Chromatogr. Sci.* **28** (1990) 357.
20. G. Guiochon and C. L. Guillemin, Quantitative gas chromatography, *J. Chromatography Library*, Vol. 42, Elsevier, Amsterdam (1988).
21. G. Schomburg, A. Deege, J. Kohler and V. Bien-Vogelsang, Immobilisation of stationary liquids in reversed-and normal-phase liquid chromatography. *J. Chromatography* **282** (1983) 27.
22. D. R. Gere, Oligomer separations by supercritical fluid chromatography. *Application Note AN 800-3*, Hewlett-Packard Co., Avondale, PA (1983).
23. J. H. Knox, B. Kaur and G. R. Millward, Structure and performance of porous graphitic carbon in liquid chromatography. *J. Chromatogr.* **352** (1986) 3.
24. S. B. Hawthorne, Analytical scale supercritical fluid extraction. *Anal. Chem.* **62** (1990) 633A.
25. M. Saito, T. Hondo and Y. Yamauchi, Fractionation by coupled micro-supercritical fluid extraction and supercritical fluid chromatography, *Supercritical Fluid Chromatography*, ed. R. M. Smith, RSC Chemistry Monographs.
26. M. Saito, T. Hondo and M. Senda, Coupled supercritical fluid extraction-supercritical fluid chromatography. *Chromatogr. Sci.* **50** (1990) 331.
27. K. R. Jahn and B. Wenclawaik, Direct on-line coupling of small subcritical and supercritical fluid extractors with packed column supercritical fluid chromatography. *Chromatographia* **26** (1988) 345.
28. M. Saito and Y. Yamauchi, Enrichment of tocopherols in wheat germ by directly coupled supercritical fluid extraction with semi-preparative supercritical fluid chromatography. *J. Chromatogr. Sci.* **27** (1989) 79.
29. S. Mitra and N. K. Wilson, Thermal modulation interface between supercritical fluid extraction and supercritical fluid chromatography. *J. Chromatogr. Sci.* **28** (1990) 182.
30. M. Ashraf-Khorassani and J. M. Levy, Quantitative analysis of polymer additives in low density polyethylene using supercritical fluid extraction/supercritical fluid chromatography. *J. High Resoln. Chromatogr.* **13** (1990) 742.
31. M. Ashraf-Khorassani, M. L. Kumar, D. J. Koebler and G. P. Williams Evaluation of coupled supercritical fluid extraction-cryogenic collection-supercritical fluid chromatography (SFE-CC-SFC) for quantitative and qualitative analysis. *J. Chromatogr. Sci.* **28** (1990) 599.
32. M. Ashraf-Khorassani, D. S. Boyer and J. M. Levy Optimisation of experimental parameters for the determination of polymer additives using on-line SFE-SFC. *J. Chromatogr. Sci.* **29** (1991) 517.
33. M. R. Andersen, J. T. Swanson, N. L. Porter and B. E. Richter, Supercritical fluid extraction as a sample introduction method for chromatography. *J. Chromatogr. Sci.* **27** (1989) 371.
34. Q. L. Xie, K. E. Markides and M. L. Lee, Supercritical fluid extraction-supercritical fluid chromatography with fraction collection for sensitive analysis. *J. Chromatogr. Sci.* **27** (1989) 365.
35. C. F. Poole and S. K. Poole, *Chromatography Today*, Chapter 6 (1991) pp. 601–648.
36. I. L. Davis, M. W. Raynor, J. P. Kithinji, K. D. Bartle, P. T. Williams and G. E. Andrews, LC, SFE, GC, SFC interfacing. *Anal. Chem.* **60** (1988) 683A.
37. M. E. McNally and J. A. Wheller, Supercritical fluid extraction coupled with supercritical fluid chromatography for the separation of sulfonyurea herbicides and their metabolites from complex matrices. *J. Chromatogr.* **435** (1988) 63.
38. E. D. Ramsey, J. R. Perkins, D. E. Games and J. R. Startin, Analysis of drug residues in tissue by combined supercritical fluid extraction-supercritical fluid chromatography-mass spectrometry-mass spectrometry. *J. Chromatogr.* **464** (1989) 353.
39. Y. Nishikawa, Application of supercritical fluid extraction/supercritical fluid chromatography to pesticide residue analysis. *Anal. Sciences* **7** (1991) 567.
40. J. Hedrick and L. T. Taylor, Quantitative supercritical fluid extraction/supercritical fluid chromatography of a phosphonate from aqueous media. *Anal. Chem.* **61** (1989) 1986.

41. D. Thiebaut, J-P. Chervet, R. W. Vannoort, G. J. De Jong, U. A. Brinkman and R. W. Frei, Supercritical-fluid extraction of aqueous samples and on-line coupling to supercritical-fluid chromatography. *J. Chromatogr.* **477** (1989) 151.
42. M. Saito and Y. Yamauchi, *Handbook, Supercritical Fluid Extraction and Supercritical Fluid Chromatography for Preparative Separation*, JASCO Report Publisher, Tokyo, Japan (1990).
43. H. Engelhardt and A. Gross, On-line extraction and separation by supercritical chromatography with packed columns. *J. High Resolution Chromatogr. Chromatogr. Commun.* **11** (1988) 38.
44. K. Sugiyama, T. Shiokawa and T. Moriya, Application of supercritical fluid chromatography and supercritical fluid extraction to the measurement of hydroperoxides in foods. *J. Chromatogr.* **515** (1990) 555.
45. Y. Ikushima, N. Saito, K. Hatakeda, S. Ito and T. Goto, Development of extraction and separation monitoring in situ with on-line SFE-SFC/FTIR system. *Chemistry Letters* **10** (1989) 1707.
46. D. A. Roston, Supercritical fluid extraction-supercritical fluid chromatography for analysis of a prostaglandin: HPMC dispersion. *Drug Development and Industrial Pharmacy* **18** (2) (1992) 245.
47. J. S. Warner and M. F. Yancey, Application of supercritical fluid extraction and supercritical fluid chromatography to residue analysis. *Brighton Crop Protection Conference-Weeds. Part 1* (1989) 267.
48. T. W. Ryan, S. G. Yocklovich, J. C. Watkins and E. J. Levy, Quantitative analysis of additives in polymers using coupled supercritical fluid extraction-supercritical fluid chromatography. **505** (1990) 273.

6 Supercritical fluid extraction-high performance liquid chromatography: on-line and off-line strategies

A.L. HOWARD and L.T. TAYLOR

6.1 Introduction

The applicability of coupling gas (GC) and supercritical fluid chromatography (SFC) directly to supercritical fluid extraction (SFE) is made inherently easier by the nature of the chromatographic mobile phases employed. For on-line GC analysis, the presence of the supercritical fluid (SF) carbon dioxide in the SFE/chromatographic interface is not a problem since the GC mobile phase itself is a gas. After the analyte is trapped at the head of the column and the decompressed extraction fluid is purged, thermal desorption of the analyte is begun. In SFE-SFC, the same fluid that is used for extraction is likewise used for chromatography. Even though the fluid is decompressed and exists as a gas in the interface, where extracted analyte is focused, raising the pressure and the temperature of the chromatographic mobile phase converts it into a SF once again for SFC.

A logical extension from the above hyphenated SFE/chromatography techniques would be the interface of SFE to high performance liquid chromatography (HPLC) since many extracts cannot be analysed chromatographically by GC or SFC due to extensive polarity, high molecular weight, and/or thermal lability. On-line SFE-HPLC would be particularly helpful for those samples that are (a) light and/or air sensitive, (b) at the trace level, and (c) in limited quantity. Trace analysis is particularly attractive for SFE-HPLC since quantitative transfer of all analytes extracted to the chromatographic system becomes possible.

Unfortunately, a SFE-HPLC interface is not as straightforward as it is in the case of either GC or SFC. The main problem with this interface lies in coupling a sample preparation technique where gas is produced in the interface to a chromatographic technique with a liquid mobile phase. In conventional HPLC analysis, maintaining the mobile phase as gas-free as possible is pertinent in optimum pump and detector performance. Helium sparging, sonication, evacuation and heating are all means by which removal of gas can be achieved prior to the use of the mobile phase. Pumping of the mobile phase becomes difficult and erratic when gas is present in the HPLC system's plumbing [1].

Difficulties such as irregular baseline noise and variable detector response also arise when bubbles are introduced into an ultraviolet absorbance (UV) detector which is most routinely used in HPLC. Dissolved gases have been likewise shown to quench fluorescent species, thus ruling out this type of detection [2]. However, the use of less conventional HPLC detectors such as the moving belt flame ionization detector (FID) [3,4] and the sulphur chemiluminescence detector (SCD) [5,6] are more compatible since they are flame-based and unaffected by gases in the mobile phase. In fact, gases such as CO_2 have been shown to be compatible with the SCD since it has been interfaced with SFC [7]. Unfortunately, the SCD can only be used with micro-HPLC (column i.d. $< 320\,\mu m$) which requires miniaturized equipment not commonly found in most analytical laboratories. All of the above difficulties indicate that an on-line SFE-HPLC system could be operated in a qualitative mode but the quantitative mode would be difficult due to possible pump and detector precision problems.

On-line SFE-HPLC, therefore, has not proven to be a very popular hyphenated sample preparatory/separation technique. SFE, on the other hand, has been shown to be quite effective in extracting HPLC-analysable compounds from difficult matrices. In the following section, SFE-HPLC experimentation is reviewed. Each study is reviewed singly so that comments on each author's approach can be made rather than generally reviewing the categories of instrumentation, trapping, interface, etc. In section 6.3, a variety of off-line applications of SFE with HPLC analysis are reviewed in order to show the applicability of SFE to HPLC-type samples. Some off-line analysis strategies are also pointed out and discussed.

6.2 On-line studies

The first report of an on-line SFE-HPLC system (Figure 6.1) was made by Unger and Roumeliotis [8] in 1983. Both static and dynamic SFE were carried out using supercritical CO_2 at constant pressure (96 bar) and temperature (40°C). The interface between the extractor and the HPLC was accomplished by means of two high pressure injection valves connected in parallel. Back-pressure in the extraction portion of the system was maintained by the use of a back-pressure regulator before the extraction vessel and a tightly packed microbore column after the first switching valve. The use of a restrictor for the SFE system was not mentioned.

The system had three distinct phases: release, load, and inject. In the release phase the first valve was switched so that the SF was vented to waste during extraction. The second valve was set so that the HPLC column was being conditioned with mobile phase. In the load position, the valves were switched in such a way that the analytes were trapped on two microbore ODS HPLC columns connected via a tee. Once the two microbore trapping columns were

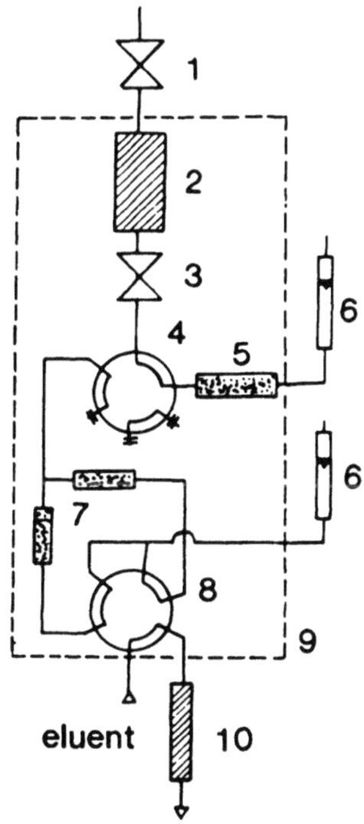

Figure 6.1 On-line SFE/HPLC system from Unger and Roumeliotis [8]. Components are as follows: (1) back pressure regulator; (2) extraction vessel; (3) high-pressure two-way valve; (4) six-port external sample valve; (5) packed microbore column for release and waste deposit; (6) rotameter; (7) microbore trap columns; (8) sample injector; (9) thermostat; (10) HPLC column.

loaded with extracted analyte, the second valve was switched to the chromatography or inject mode. In this mode the HPLC mobile phase was directed into these traps in order to carry the analyte(s) to the chromatographic analytical scale column which was being conditioned while the extraction was taking place. Only normal phase HPLC with UV detection was investigated. The sample studied was the extraction of valepotriates from *Radix valerianae*.

The assessment of this extraction/chromatography system was difficult for several reasons. In any HPLC system, dead volume in the injection area becomes critical due to its band broadening effect. The band broadening associated with using two microbore columns and a tee as a trap and pseudo injection loop could possibly introduce enough dead volume to cause extensive peak broadening. However, the assessment of peak shape or chromatographic integrity in general was not possible since no chromatograms were shown in the article. Detector stability via baseline noise examination could not be done either. The only semi-quantitative data that were given consisted of a kinetics or time study plot for the extraction of valtrate from *Radix valerianae* (Figure 6.2). Since the units of recovery for this analyte were given as relative mass units/time, no quantitative conclusions could be drawn.

other words when an exhaustive extraction of the analytes had occurred, the switching valve injected 10 or 20 µl of the supercritical CO_2 stream into the HPLC system.

The chromatographic system employed with the SPA utilized a reversed phase mobile phase (acetonitrile/water) and gradient elution. Analyte detection was done by ultraviolet absorbance at 254 nm. The extraction of caffeine out of instant coffee [10] (Figure 6.4) and ibuprofen and caffeine out of a crushed ibuprofen tablet [10] (Figure 6.5) were given as applications. It should be noted that both chromatograms have an interesting peak that is labelled CO_2, indicating that it apparently elutes as a plug rather than diffusing into the mobile phase.

Obviously this SFE-HPLC system was designed for analytes that are easily analysed by UV. In the case of the SFE-HPLC system, this is not a problem since UV is also used as the HPLC detector. The manufacturer claimed that the SPA was also easily coupled to GC or SFC. Analytes that are amenable to GC analysis often do not contain UV chromophores. Even if the analyte is sensitive in the UV, it may not be very sensitively analysed by the detector used with a GC or SFC assay. Likewise, a weakly UV absorbing substance that is detectable in the UV may overload a highly sensitive or selective GC or SFC detector. As an example, consider a polychlorinated biphenyl (PCB) as a UV-absorbing analyte commonly assayed by GC. If analysis were done by GC-FID, the analyte response would probably be much lower than that of

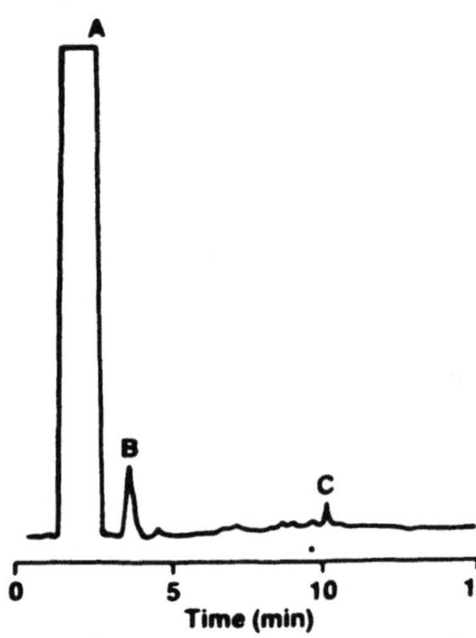

Figure 6.4 On-line SFE-HPLC chromatogram of an instant coffee extract from Nair and Huber [10]. Peaks are as follows: (A) CO_2; (B) caffeine; (C) unknown. Conditions: SF, CO_2; pressure, 5000 psi; temperature, 50°C; mobile phase, (10% acetonitrile/90% water)/acetonitrile; flow rate, 1 ml/min; column, 10 cm × 4.6 mm i.d. C18 column; detection, UV at 254 nm. Analyte structures given in Table 6.1.

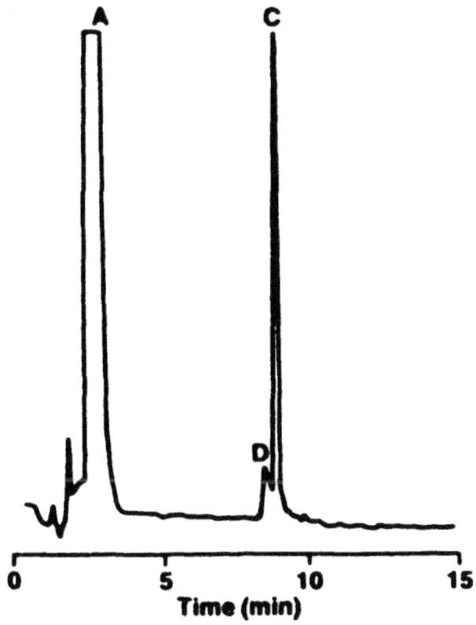

Figure 6.5 On-line SFE-HPLC chromatogram of a crushed ibuprofen tablet extract from Nair and Huber [10]. Peaks are as follows: (A) CO_2; (C) ibuprofen; (D) unknown. Conditions same as in Figure 6.4. Analyte structures given in Table 6.1.

the UV sensor due to the quenching effect that heteroatoms have on FID response. On the other hand, if GC-electron capture detection (ECD) were used, the detector would most likely be overloaded due to the fact that the ECD is orders of magnitude more sensitive to chlorinated compounds than the UV. Consequently, because the UV detector used as an extraction sensor is not universely or uniformly sensitive to all analytes, one must consider its use carefully when interfacing to other chromatographic techniques.

Another concern with the SPA system may be the mode in which the recirculating supercritical CO_2 analyte is introduced into the HPLC system. Only a 10 or 20 µl CO_2 plug [10] is injected into the HPLC. This may not be the best scenario if trace analysis is desired of a weak UV chromophore. A SFE-HPLC system such as the one proposed by Unger and Roumeliotis [8] would probably be more appropriate for this situation because all extracted analytes deposited onto the trap are injected into the HPLC system. For analyses of bulk and/or highly UV absorbing analytes, the introduction of only a small amount of analyte may, on the other hand, be beneficial.

A third concern with the SPA system is that it only accommodates static extraction. Dynamic extraction, where fresh SF is constantly in contact with the sample, may be more appropriate since dynamic extractions are by nature exhaustive where static extractions may or may not be depending on how much analyte is present and its solubility in the SF.

The most recent on-line SFE-HPLC work reported was done by Daimon *et al.* [11] and was presented at the Twelfth International Symposium on Capillary Chromatography in 1990. The system used by the authors was modified

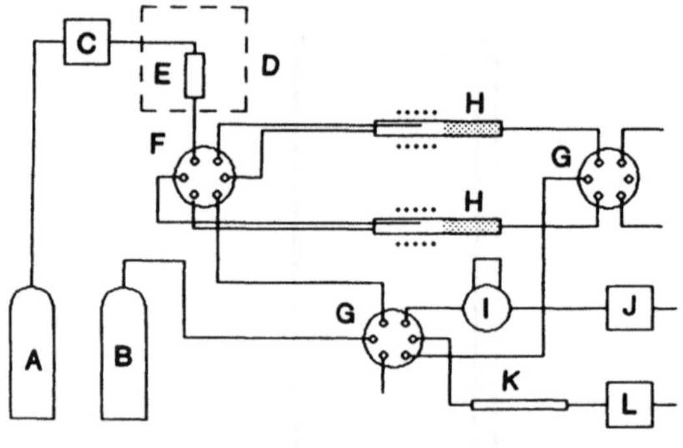

Figure 6.6 On-line micro-SFE-HPLC system from Daimon *et al.* [11]. Components are as follows: (A) CO_2; (B) N_2; (C) SFE pump; (D) oven; (E) extraction chamber; (F) double six way valve; (G) six way valves; (H) traps; (I) injector; (J) LC pump; (K) LC column; (L) UV detector.

in varying ways in order to accommodate the following on-line techniques: SFE-SFC, SFE-HPLC and HPLC-SFC. The SFE-HPLC system (Figure 6.6) employed micro-apparatus (i.e. sample weight < 0.5 mg, HPLC column 530 μm i.d. and mobile phase flow rate 2 μl/min). It is of similar design to that of Unger and Roumeliotis [8] with a few improvements. Like Unger's SFE-HPLC system, this on-line system has two microbore columns available for trapping (2 cm × 0.7 mm i.d.) but only one column at a time is used. This allows another extraction to take place while the HPLC analysis of the previous sample is being performed, thus reducing analysis time. Another improvement is the availability of three different chromatographic sample introduction modes. Analyte introduction can be carried out by using either an injector which bypasses the traps, an injection through the trap, or an SFE injection similar to that of the SPA interface. These three injection modes were compared (Figure 6.7) by the extraction of three polymer additives out of a polypropylene sample. As can be seen, the peak shapes obtained for the

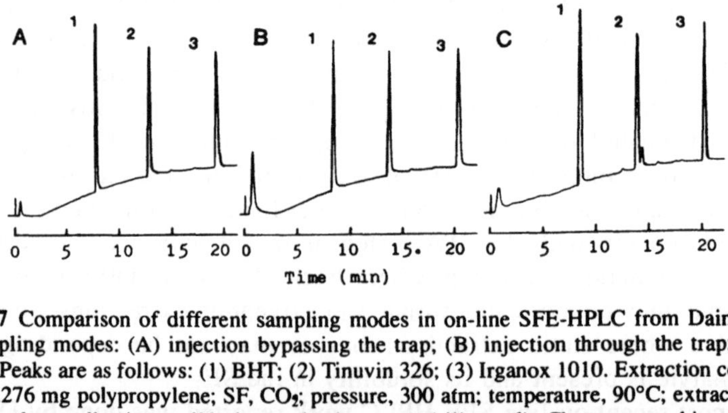

Figure 6.7 Comparison of different sampling modes in on-line SFE-HPLC from Daimon *et al.* [11]. Sampling modes: (A) injection bypassing the trap; (B) injection through the trap; (C) SFE injection. Peaks are as follows: (1) BHT; (2) Tinuvin 326; (3) Irganox 1010. Extraction conditions: sample, 0.276 mg polypropylene; SF, CO_2; pressure, 300 atm; temperature, 90°C; extraction time, 30 min; analyte collection, solid phase adsorbent trap (silica gel). Chromatographic conditions: column, 15 cm × 0.53 mm i.d. Develosil ODS-5K; mobile phase, gradient programmed from 50% to 100% acetonitrile in water; detection, UV at 210 nm. Analyte structures given in Table 6.1.

through-the-trap injection and the SFE injection are not significantly broader than those using the bypass injection except in the case of peak 2 in chromatogram C which appears to be split. This preservation of chromatographic integrity indicates that the authors were quite effective in reducing system dead volume to a minimum. This is of particular concern in a micro-system where dead volume is an order of magnitude more important than it is in an analytical scale system. An additional advantage to having the bypass injector is that the system can also be used as a stand-alone micro-HPLC without any further equipment modification. Likewise, this feature could also be an advantage for the SPA interface if the switching valve were replaced with a four or six port valve. Unger's system would require more extensive modification in order to achieve this advantage.

Another strong point of this system is the use of gradient elution for rinsing analytes off the normal phase microbore trap. The reversed phase solvent strength during the rinsing of this trap is kept low so that all analytes can be concentrated at the head of the reversed phase HPLC chromatographic column before the gradient ramp begins, thus making HPLC peak shape more symmetrical. An exponential type gradient profile proved to be the most effective. It has been established that generating a reproducible gradient profile in micro-HPLC [12] is more difficult than it is in analytical scale HPLC even if syringe pumps are used.

Another more general disadvantage of this SFE-HPLC system is that it is a micro-system which requires the extraction of very small sample sizes in order that extracts will not overload the HPLC system. Sample sizes used in this system were typically < 0.5 mg polypropylene. For these types of sample sizes, the question of sample homogeneity becomes relevant. This is particularly a problem when comparing recovery and reproduciblity data to those obtained by sample preparation techniques, such as Soxhlet or liquid extractions, where the sample sizes used can be orders of magnitude larger. In short, it is important to realize that differences in these data may not be due to poorer extraction recovery or precision but to sample size. We have found this to be the case in our laboratory [13] in the off-line SFE-HPLC of a hypolipidemic drug from a mechanically prepared rat feed matrix.

One way to examine the effect of sample size on extraction recoveries and precision may be to extract larger sample sizes with collection off-line rather than direct, on-line analysis. This could easily be done by including a second HPLC pump in the system so that the microbore analyte trap could be rinsed into a vial off-line. With the SFE-HPLC system proposed by Daimon *et al.* [11], a separate HPLC would not be necessary due to the bypass injector with which the system is equipped. In this way, the more concentrated extract could be analysed off-line with the same HPLC system where precautions against sample column overloading can be exercised.

Column overload even with small sample sizes may be a problem with a micro-HPLC system. This is particularly true where the extract is injected via the through-the-trap method since all analyte extracted is placed on the HPLC system. For extracts which have high analyte concentrations, the SFE injection procedure would be more applicable since it offers the capacity to limit the amount of sample injected. Again the capacity for different modes of injection is the key to this system's success.

6.3 Off-line studies

We have included this section in the chapter in order to illustrate that even though on-line SFE-HPLC has not been quantitatively feasible, SFE is quite useful for quantitative determination of those analytes that must be analysed by HPLC and should not be ruled out when considering sample preparatory techniques.

Work reported by Hirata and Okamato [14] in 1989 illustrates a type of compromise between on- and off-line HPLC analysis. In this investigation, the authors used a microbore column packed with silica as a trap similar to that used by Daimon *et al.* [11]. Instead of rinsing the trap off-line with solvent or connecting the trap on-line to the HPLC, the trap was removed from the extractor and placed in the HPLC system just before the column. In this capacity, it is used as an injection loop. The use of this trap as the injector is beneficial for several reasons. First, it allows all extracted components to be chromatographed directly as in an on-line system. It also avoids the rinsing step needed when analytes are deposited on a solid phase sorbent after SFE. This rinsing feature adds another step to the sample preparatory process which could result in higher inherent error in the method. In short, using this rinsing procedure simultaneously as the injection step garners some on-line advantages while at the same time reducing inherent off-line experimental error. Conversely, there are disadvantages to this type of injection such as column overload.

An interesting comparison was drawn between a Soxhlet and a SFE for the extraction of polyethylene film. Both analyses were done by reversed phase HPLC with UV detection. The Soxhlet extraction was performed for 2 days with dichloromethane while the SFE was performed using CO_2 at 35°C and 250 atm for 2 h. Both employed a polyethylene sample that weighed approximately 3 mg. As can be seen in Figure 6.8, the chromatograms appear to be quite similar in the types of peaks present but dissimilar in their intensity. It is rather interesting though that extractions using solvents of such different polarity and extraction time could produce chromatograms so similar. A significant advantage of SFE is realized in this comparison in that it could replace the use of chlorinated solvents for extraction without sacrificing any extraction effectiveness.

A sample size problem was pointed out when a SFE reproducibility study was performed on five 3-mg samples of polyethylene. Several HPLC chroma-

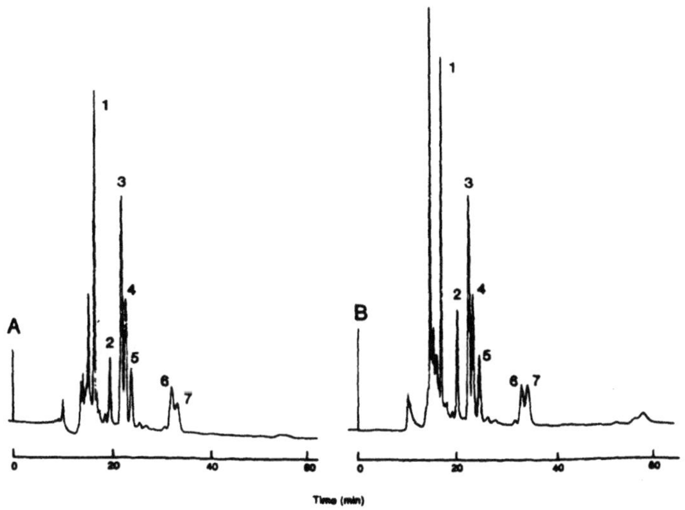

Figure 6.8 Comparison of (A) Soxhlet extraction* and (B) SFE of polyethylene samples from Hirata and Okimato [14]. Soxhlet conditions: sample size, 10.65 g; solvent, dichloromethane; extraction time, 2 days; extract volume, 10 ml. SFE conditions: sample size, 3.024 mg; SF, CO_2; pressure, 250 atm; temperature, 35°C; extraction time, 2 h; analyte collection, solid phase adsorbent trap (silica gel). Chromatographic conditions: column, 55 cm × 0.53 mm Kaseisorb ODS-5; mobile phase, methanol; injection volume*, 3 µl of stock Soxhlet extract.

tographic peaks were used for comparison. Depending on which peak was used, the recovery RSDs ranged from 4 to 29%. This broad range of RSDs was found by a process of elimination to be due to sample inhomogeneity. Figure 6.9 demonstrates this phenomenon in that when different size samples

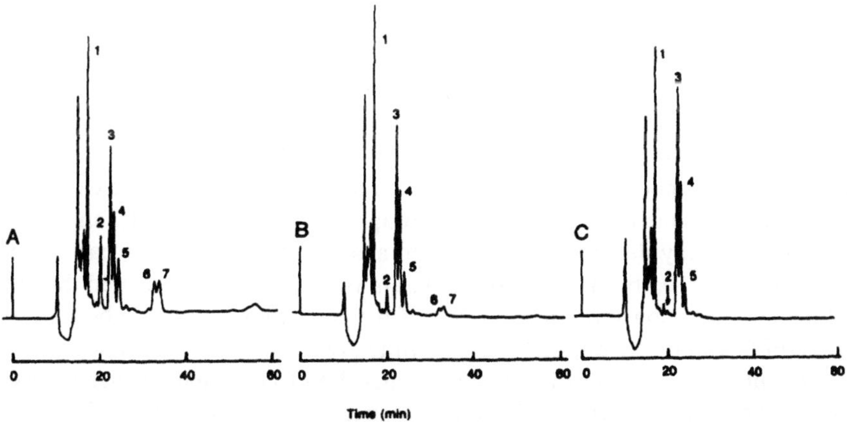

Figure 6.9 Inhomogeneous distribution of additives in (A) 0.728 mg, (B) 0.580 mg, and (C) 0.494 mg polyethylene samples determined by SFE from Hirata and Okimato [14]. SFE and HPLC conditions as in Figure 6.8.

were extracted under identical conditions, the peaks obtained for the extract of the 0.580 mg sample were more intense than those from the larger 0.728 mg sample. This method of extraction and chromatography is probably not appropriate for this type and size of sample. However, this system could have application for those samples that are of a restricted size by nature such as biological ones.

Sample preparatory procedures in environmental analysis are of particular interest because they are often rather cumbersome and time-consuming. An application of off-line SFE with HPLC analysis is the quantitative extraction of anthraquinone out of paper and wood by Schneiderman *et al.* [15]. Anthraquinone is used as a catalyst in order to aid delignification and to increase yield in the paper manufacturing industry [16]. It is currently determined by a variety of methods including Soxhlet extraction with HPLC-UV analysis. The purpose of this study was to see if SFE had any advantages over Soxhlet extraction for the analysis of this sample.

Static SFE was performed using CO_2 at 8000 psi and 65°C on samples of Kraft paper and pine plywood sawdust. The HPLC assay was reversed phase and employed an electrochemical detector (ECD). The ECD was used in an attempt to gain more selective and sensitive detection of the anthraquinone over other co-extractives. Supercritical fluid extraction recovery and precision data were obtained by spiking the sawdust (500 mg) or paper (300 mg) with varying amounts (3–500 μg) of anthraquinone. Quantitative (>95%) SFE of the anthraquinone was achieved in 20 min for most concentrations of the analyte with a RSD of under 3% ($n = 3$). Difficulties occurred at the lower concentrations, particularly in the case of the sawdust matrix where irreversible adsorption of the analyte to the matrix was probably occurring. The recoveries obtained with acetonitrile or methanol Soxhlet extractions were not quantitative (82%) with extraction and analysis per sample taking a total of 4 days to complete.

The quantitation of pesticides out of sediment is yet another application where off-line SFE-HPLC was found to be applicable. Janda *et al.* [17] reported a study where five s-triazines were quantitatively extracted from a 500 mg sediment sample in 30 min with methanol-modified CO_2 at 230 atm and 48°C. The modifier (200 μl) was added to the sediment prior to extraction. The extract was then collected in a liquid trap. Both capillary GC-FID and HPLC-photodiode array detection were used to analyse the extracts. As can be seen from Figure 6.10, the HPLC chromatogram is much less complicated than that of the GC. Low concentration extracts of the pesticides could only be analysed via HPLC due to the presence of co-extractives in the GC chromatogram.

This illustrates one of the advantages that HPLC-UV analysis has over GC for the analysis of extracts. For many matrices such as sediment, CO_2 is not a very selective SF and, as a result, many matrix components are co-extracted with the analyte(s). In most cases, these co-extractives are present in a GC chromatogram and not in the HPLC trace due to the fact that most of these

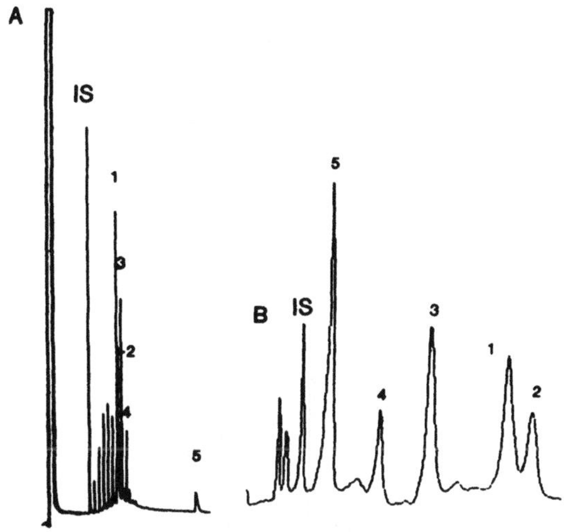

Figure 6.10 Analysis of sediment extracts by (A) capillary GC and (B) HPLC from Janda *et al.* [17]. Peaks are as follows: (IS) eicosane (GC) and theophylline (HPLC), (1) propazine, (2) terbutylazine, (3) atrazine, (4) simazine, (5) cyanazine. SFE conditions: sample size, 500 mg; SF, methanol-modified CO_2; pressure, 230 bar; temperature, 48°C; extraction time, 30 min; analyte collection, solvent (methanol). GC conditions: carrier gas, H_2; column, 30 m × 0.3 mm i.d. Superox 20M; injection volume, 2 μl; detection, FID. HPLC conditions: column, 25 cm × 4.6 mm i.d. reversed phase; mobile phase 65 methanol/35 water; flow rate, 1 ml/min; injection volume, 10–25 μl; detection, photodiode array UV at 225 nm. Analyte structures given in Table 6.1.

compounds do not contain a UV chromophore. This can make the extract chromatogram much easier to interpret. No significant time sacrifice is made since the GC method takes 13.67 min versus 16 min when analysed by HPLC.

The extraction of pesticides out of aqueous matrices is also of great interest. Classical sample preparation techniques for aqueous matrices are quite difficult due to the complexity of the matrix. Unfortunately, direct SFE of aqueous samples is also not feasible. In response to this dilemma, a method was devised by Howard and Taylor [18] to extract two sulphonyl urea herbicides, sulphometuron methyl and chlorsulphuron, out of water using SFE. This system involved the solid phase disk deposition (SPEDD) of the analytes from an aqueous sample via a vacuum. These disks are made of two Teflon™ membranes with packing material (C18, C8 cyclohexyl) suspended between them in a Teflon™ web. After deposition, the disc was placed in an extraction vessel and extracted with 2% methanol-modified CO_2. The extract analysis was then accomplished by reversed phase HPLC-UV.

Analyte trapping on the SPE disk and SFE recovery from the disc were found to be pH dependent. The optimum trapping was obtained with a water sample of pH 3. Unfortunately, the SFE was not quantitative at this pH due to strong analyte interaction with the active sites of the disc's packing material (C18). It was discovered if 2 ml of water (pH 9) was passed through the disc after trapping (at pH 3), quantitative SFE could be indeed achieved with a

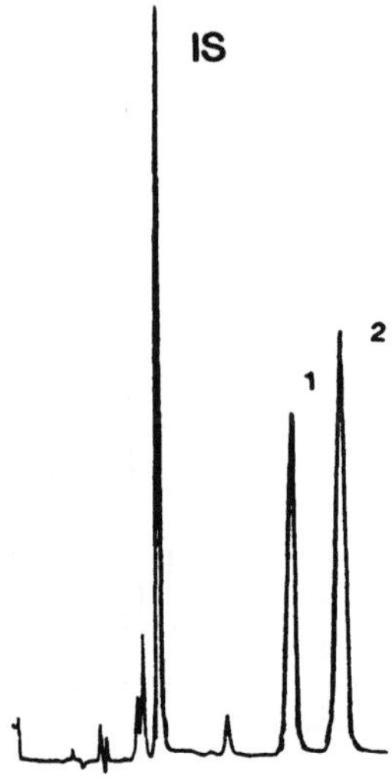

Figure 6.11 HPLC-UV chromatogram of an aqueous sample extract by SPEDD-SFE. Peaks are as follows: (IS) sulphachloropyridazine, (1) sulphometuron methyl, (2) chlorsulphuron. SPEDD conditions: sample size, 1 l water spiked with 50 μg each component; analyte deposition, pH 3 with a 2 ml disk wash of pH 9 water. SFE conditions: SF, 2% methanol modified CO_2; pressure, 350 bar; temperature, 50°C; extraction time, 24 min; flow rate, 2 ml/min.; extraction time, 24 min; analyte collection, solid phase adsorbent trap (stainless steel); rinse solvent, acetonitrile; trap temperature, 15°C during extraction, 60°C during rinsing. Chromatographic conditions: column, 25 cm × 4.6 mm i.d. C_8; mobile phase, 25 ACN/75 water pH 3; flow rate, 1 ml/min; injection volume, 20 μl; detection, UV at 230 nm. Analyte structures given in Table 6.1.

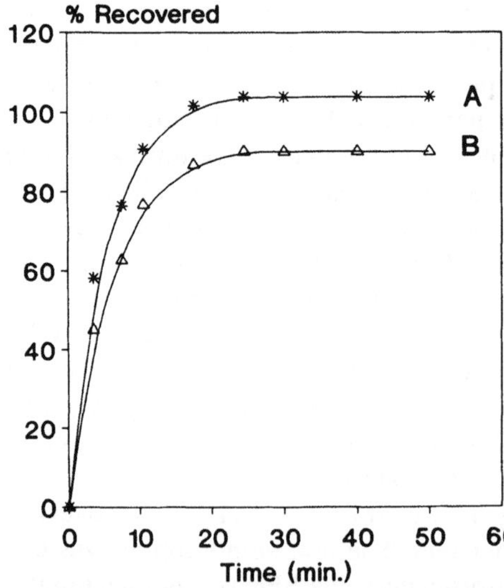

Figure 6.12 Time study profile for the extraction of (A) sulphometuron methyl and (B) using the SPEDD-SFE sample preparation system. Conditions as in Figure 6.11 except that SPEDD done at pH 7. Low chlorsulphuron recovery is due to trapping pH. Analyte structures given in Table 6.1.

RSD of < 5% ($n = 5$). No interferences (Figure 6.11) were seen in the UV chromatogram. Sulphachloropyridazine was used as an internal standard. The time study profile (Figure 6.12) for the extraction of both analytes with the SPEDD-SFE system indicates nearly instantaneous extraction. This was expected due to the analytes being extremely soluble in the modified fluid and the porosity of the SPE disk.

Off-line SFE-HPLC has also found application in the food industry in the decaffeination of beverages such as coffee and soft drinks. Work reported by Ndiomu and Simpson [19] compares the use of SFE, high pressure Soxhlet (HPS), and traditional Soxhlet extraction for the quantification of caffeine out of kola nuts which are used in the manufacture of cola-type beverages. Extract analysis was by reversed phase HPLC with UV detection in all cases. The HPLC was carried out with either methanol or tetrahydrofuran at subcritical conditions. Of these three sample preparation techniques, SFE had the worst caffeine recovery after a 3 h extraction with CO_2 at 1050 psi and 60°C. HPS using methanol proved to be the best technique since recoveries of 99.9% were achieved with only a 30 min extraction. Low SFE recoveries could be attributed to two factors. First, the extraction was carried out at a comparatively low pressure although the system was operated at its maximum operating pressure. Second, the kola nuts were dried before extraction.

Several of the problems encountered by Ndiomu and Simpson [19] were coincidentally investigated by Sugiyama *et al.* in their SFE of caffeine from ground processed coffee beans [20]. In the coffee industry, caffeine is commonly extracted from the green beans which have a particular water content [21]. Since the investigators were using dried, processed beans they added varying amounts of water to the ground sample in order to mimic the green bean extraction. Extraction parameters of pressure, temperature, time and amount water added to the ground coffee were investigated in detail. Optimum extraction conditions were found when CO_2 conditions were at 250 bar, 48°C, 60 min extraction time and 20% water added. The temperature of extraction and the amount of water added to the coffee sample appeared to have the most drastic effects on recovery. Tremendous decreases in caffeine recovery occurred when the temperature was increased at a fixed pressure. This is probably due to the fact that the SF density decreased. The same effect was observed when the water added to the coffee was reduced.

The studies of Sugiyama *et al.* [20] and Ndiomu and Simpson [19] illustrate the point of fully investigating and optimizing any extraction procedure before comparisons are made to other methods. For example, if one were to have only examined the work of the latter authors, one might determine that SFE was not a suitable sample preparation technique for caffeine due to the low recoveries. In reality, the method was somewhat biased because it was not optimized before comparisons to more successful methods were made. In short, it is important when examining the SFE literature, to look for how the authors arrived at the SFE method they use because an unoptimized extraction procedure may lead one to rule out SFE as an appropriate sample preparation technique for a particular analyte and/or matrix.

The determination of vitamin K_1 is another successful application of off-line SFE-HPLC in the food industry. A study performed by Schneiderman *et al.* [22] detailed a one-step SFE method for the extraction of vitamin K_1 out of both soy protein and milk-based infant formulas. Even though the amount of all vitamins in infant formulas is strictly controlled and tested, there is no standard method of analysis to date. The authors use the same extraction and chromatographic system as in reference [15] except for the addition of methylene chloride to the reversed phase mobile phase in order to improve peak shape.

Before extracting the vitamin K_1 out of the infant formulas, the authors first determined the extractability of the analyte alone with CO_2 at 8000 psi and 60°C. A 92% recovery with a RSD of 3% ($n = 7$) was obtained under these conditions for an extraction time of 15 min. Varying sample sizes of the powdered infant formulas were then spiked with varying amounts of vitamin K_1 and extracted under these conditions. Recovery and reproducibilty data for the vitamin from the soy protein and milk-based infant formulas were 94.4% and 95.6%, respectively. Relative standard deviations were 6.5% and 7.4%, respectively. Actual samples of unspiked infant formulas where the amount of vitamin K_1 was known to be 0.8 μg/g were also assayed. The SFE method

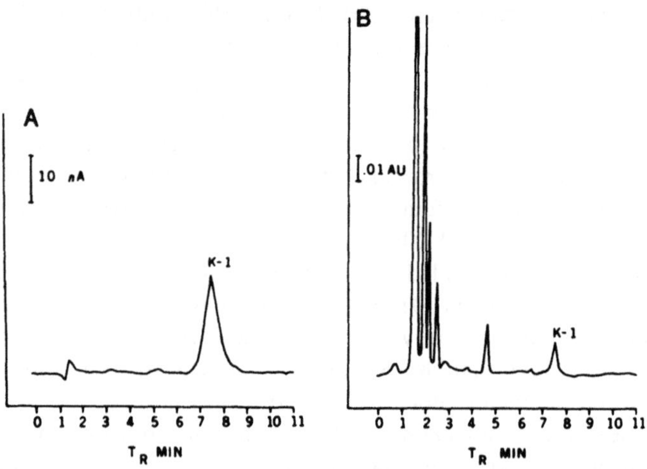

Figure 6.13 SFE extract analysis of vitamin K_1 in (A) milk-based formulas by HPLC-ECD and (B) soy protein-based formula infant formulas by (B) HPLC-UV from Schneiderman *et al.* [22]. Extraction conditions: sample size, 1 g; SF, CO_2; pressure 8000 psi; temperature, 60°C; extraction time, 15 min; analyte collection, solid phase adsorbent (silica gel); rinse solvent, 50:50 methylene chloride/acetone. Chromatogram A conditions: column, 15 cm × 3.9 mm C_{18}; mobile phase, 90 ACN/5 CH_2Cl_2/5% 0.025 M $NaClO_4$ in water; flow rate, 2 ml/min; injection volume, 20 μm, detection, electrochemical. Chromatogram B conditions: mobile phase, 90:5:5 water; detection, UV at 248 nm; others as in A. Analyte structures given in Table 6.1.

found the levels to be 0.87 and 0.95 μg/g, respectively, indicating that there may be more vitamin K_1 present in the 'real' sample than was thought. As can be seen in Figure 6.13, there were indeed components of the formulas extracted but because they are not electroactive, no response was seen with the electrochemical detector.

The quantitation of menadione (vitamin K_3) out of animal feed using off-line SFE-HPLC-ECD by Schneiderman *et al.* [23] likewise illustrates the determination of a lipophilic analyte from a food-type matrix. Menandione is a synthetic pro-vitamin that is converted into the active vitamin K_2 in the liver. It is added to animal feeds in order to combat blood clotting deficiencies in animals brought about by antibiotics or anticoagulant medications. Quantitative SFE recovery of menadione was achieved from spiked rat feed (0.2 mg menadione/500 mg) samples in 20 min with SF CO_2. Low recoveries below this spiking level were determined to be a result of irreversible absorption of the vitamin onto the matrix. Possibly the use of a modifier in the SF would allow spiking levels below this to be examined.

The pharmaceutical industry represents yet another area where off-line SFE-HPLC has found application. Dosage uniformity and potency verification as well as toxicological and clinical studies are all applicable to this technique. As an example, a study by Messer and Taylor [13] was performed in order to determine the dosage uniformity of a hypolipidimic drug in a rat feed matrix for the purpose of toxicological studies. The first step in this study was to determine the drug's extractability under varying extraction conditions in order to optimize the method. The extracts were analysed by reversed phase HPLC/UV.

After determining the optimum extraction method, 100 mg of the original drug-containing rat food sample (1% w/w of the drug) was extracted. Less than quantitative recovery and poor reproducibility were the result. By preparing several different samples where the drug was added either as a solution or ground along with the rat food, it was determined that the original rat feed matrix was not very homogeneous. The original sample was prepared by mixing the granular drug vigorously with the ground rat feed in a large mixer. The sample size of the original sample employed with the currently used method of sample preparation was 3 g which would be inherently more representative. The rat feed sample where the drug was added as a solution, allowed to evaporate, and then vigorously mixed proved to be the most homogenous of those investigated. For samples such as this one, where sample inhomogeneity may be a problem, on-line SFE-HPLC could not even be considered due to the sample size restrictions posed by such a system.

Another example of the use of SFE-HPLC in the pharmaceutical industry is the verification of dosage in a pharmaceutical formulation [24]. The active components, sulphamethoxazole and trimethoprim, of the intravenous drug Septra Infusion were successfully extracted from the excipients and chromatographed by reversed phase HPLC-UV. The active components in the preparation were at rather high concentrations, 80 and 16 mg/ml, respectively. Attempts were made to extract the intravenous solution directly [25] with little success due to restrictor plugging. High analyte concentration in the formulation and the precipitation of sulphamethoxazole out of the Septra solution as CO_2 was bubbled through the solution, both contributed to this problem. The precipitation was caused by a decrease in the formulation's pH from 10 to 3 as the CO_2 was passed through. It was determined that sulphamethoxazole was soluble in the preparation level concentrations at a pH > 10.

All of these problems were avoided by spiking 100 µl of a diluted solution of the Septra Infusion onto celite, allowing the solvent to evaporate, and extracting with CO_2. Quantitative extraction of sulphamethoxazole was achieved with an RSD < 2% but trimethoprim was only recovered at the 86% level with an RSD < 5%. A time study is shown in Figure 6.14. Due to the shape of these curves, it can be seen that the extraction is diffusion limited for both analytes for most of the extraction.

As was described previously, SFE was not amenable for the extraction of the liquid sample. The use of the celite illustrates another use of a solid support; its ability to immobilize liquid samples. For the Septra formulation, this system was ideal since the intravenous solution had high analyte concentrations so only small amounts of liquid needed to be added to the celite. In the case of more dilute liquid samples, the use of a solid support for this

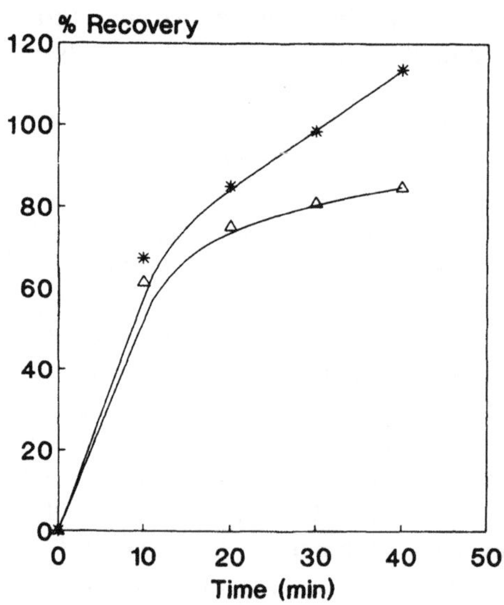

Figure 6.14 SFE time study for the extraction of (A) sulphamethoxazole and (B) trimethoprim out of Septra Infusion drug formulation deposited on celite. Extraction conditions: sample size, 100 µl; SF, CO_2; pressure, 329 bar; temperature, 60°C; trap temperature, 5°C; trap rinsed with HPLC mobile phase (acetonitrile/water) at four 10-min intervals. Analyte structures given in Table 6.1.

purpose may not be appropriate because of the excessive amounts of liquid needed to be placed on the support to allow for an assayable analyte concentration in the extract. Analyte breakthrough in this scenario could be a large source of error during the solvent evaporation step if the celite is overloaded.

6.4 Conclusions

The analysis of extracts by HPLC has one inherent advantage over GC whether the analysis be off- or on-line. The extract chromatogram is usually much simpler than that of a GC analysis since most co-extractives do not contain UV chromophores. Most organic compounds do, however, respond in a FID. As a result quantitation from a UV chromatogram may involve less error due to more reliable integration. Unfortunately in most cases, HPLC does involve longer analysis times than both GC and SFC.

At present, on-line SFE-HPLC appears to be feasible for qualitative analysis only. All of the systems detailed above suffer from disadvantages, such as gas introduction in the mobile phase, which preclude their use in accurate, precise analytical determinations. The most promising design of all those mentioned is the system proposed by Daimon *et al.* [11] because of its versatility concerning sample introduction into the HPLC. Even this system, however, has a major drawback in that it is a micro-system.

Of the systems examined, only those that use solid sorbent traps allow for the use of dynamic as well as static SFE while those where a CO_2 plug is injected can only be used in the static mode. The system by Daimon *et al.* [11] allows for both so that the system could be customized to extract samples in an individualized manner. However, both types of sample introduction procedures have their problems. It appears that rinsing of the trap is more difficult in the former where small sample injection size is the limiting factor. In the latter, the use of gradients for trap rinsing significantly improves peak shape by focusing the analytes on the head of the HPLC column prior to separation.

No attempts to remove the CO_2 left in the trap after decompression were made in any of the systems employing this type of analyte collection. At the beginning of this chapter, this was cited as a principal drawback of the system. However, after further examination of the work reviewed here, sample size restrictions also appear to be another significant barrier to using on-line SFE-HPLC for quantitative analysis of 'real' samples.

Despite all these drawbacks of the on-line technique, off-line SFE-HPLC appears to be applicable and quantitative for a variety of samples in many 'real' world matrices. In most cases, all of the disadvantages mentioned with the on-line technique are eliminated. Even extract manipulation can be minimized if the HPLC mobile phase, or a component of it, is used as a rinse solvent. Care should be taken, however, to ensure that the trapping material and the rinse solvent are compatible and that the analyte(s) are sufficiently soluble in the rinse solvent. For example, solvents such as acetonitrile or methanol should be used with reversed phase-type traps while solvents such as methylene chloride and hexane should be used with normal phase types. In most cases this translates into the ability of being able to inject the extract directly into the chromatograph and results in a reduction in experimental error. The level of RSDs quoted for all of the off-line applications are well within what one would expect for analysis of complex samples. Conversely, those quoted by Daimon *et al.* [11] were not. The cause for this large RSD was said to be sample related although extractions of no other samples were investigated to confirm it.

As can be seen from the examples (Table 6.1) given throughout this chapter, the range of HPLC-type analytes applicable to SFE is enormous. At first glance, it might appear that many of these analytes may be too polar to be extracted with pure CO_2. Only those analytes denoted with an * required the use of modified CO_2. The matrices that the analytes were removed from were also far from trivial, proving once again that SFE is indeed applicable to real world samples containing analytes that are not necessarily amenable to GC analysis.

Table 6.1 Summary of analytes determined by SFE-HPLC

Analyte	Matrix	Reference
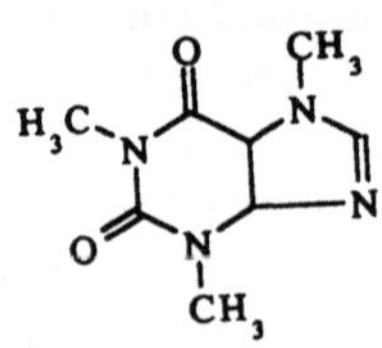 Caffeine	Instant coffee	10
	Kola nuts	19
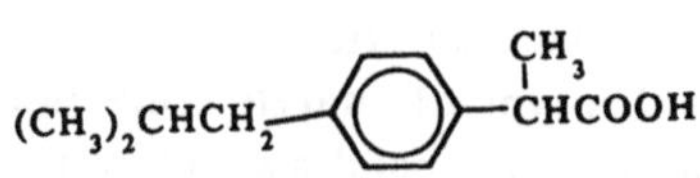 Ibuprofen	Ibuprofen tablet	10
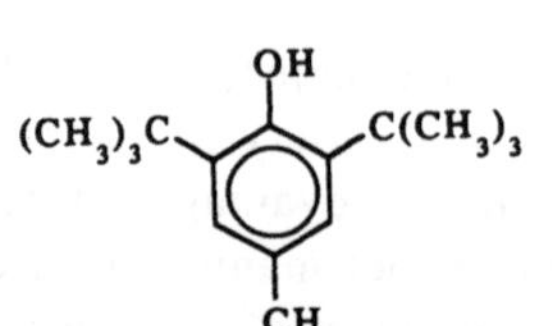 BHT	Polypropylene	11

Table 6.1 *(continued)*

Analyte	Matrix	Reference
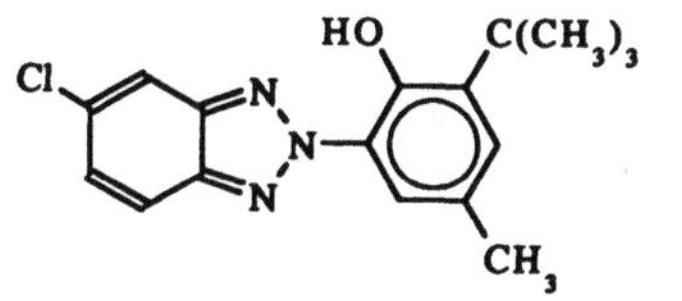 Tinuvin 326	Polypropylene	11
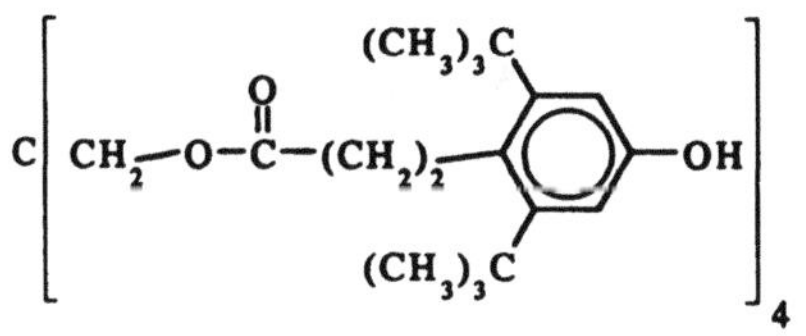 Irganox 1010	Polypropylene	11
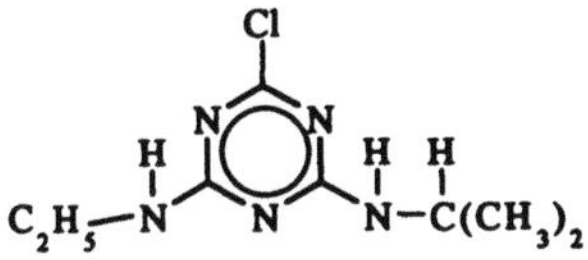 * Antrazine	Sediment	17
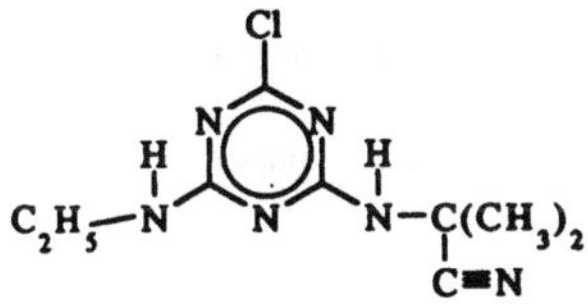 * Cyanzine	Sediment	17
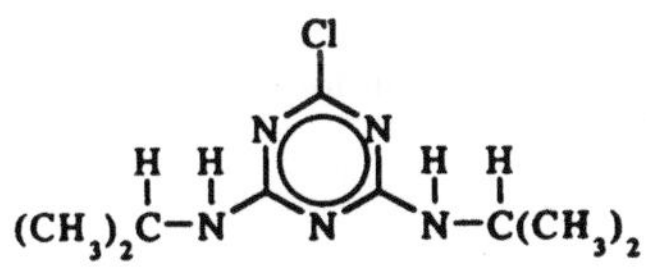 * Propazine	Sediment	17
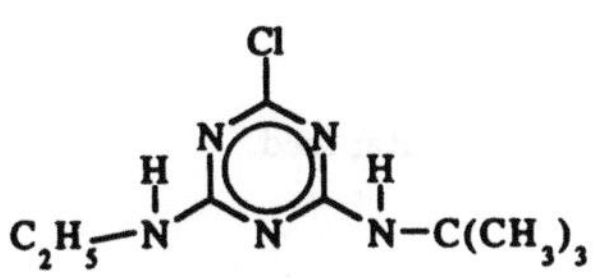 * Terbutylazine	Sediment	17

Table 6.1 *(continued)*

Analyte	Matrix	Reference
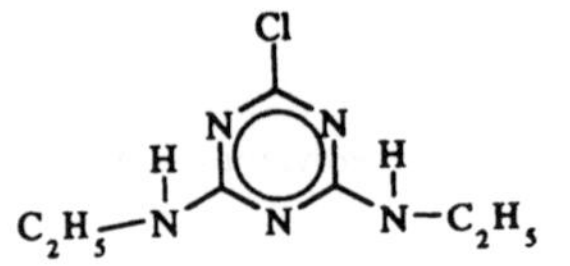 * Simazine	Sediment	17
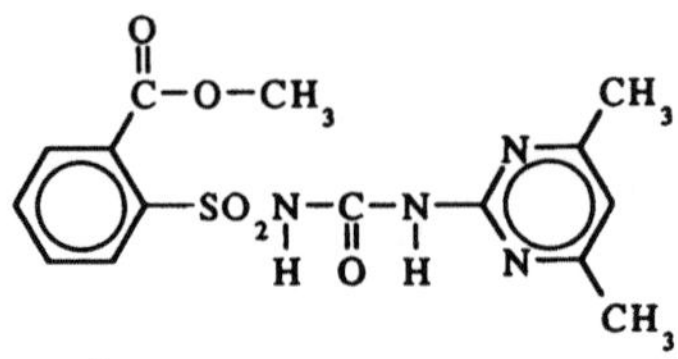 * Sulphometuron methyl	Water	18
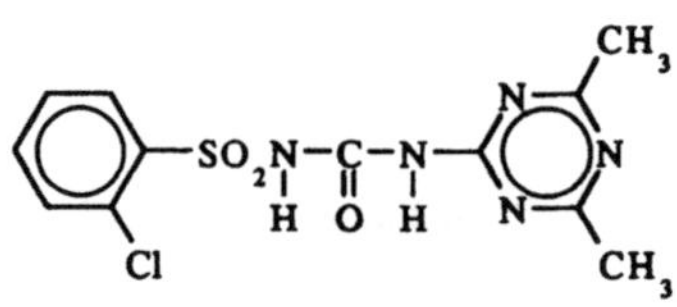 * Chlorsulphuron	Water	18
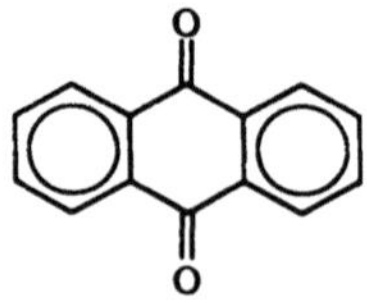 Anthraquinone	Pine sawdust Kraft paper	15
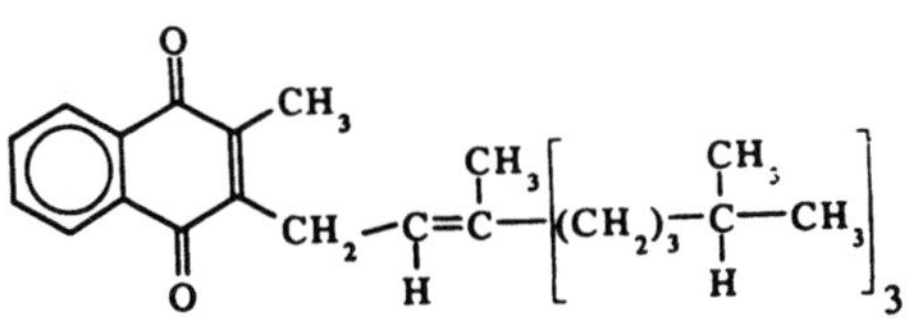 Vitamin K_1	Infant formula	22
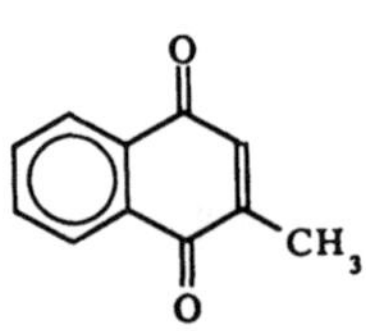 Vitamin K_3	Rat feed	23

Table 6.1 *(continued)*

Analyte	Matrix	Reference
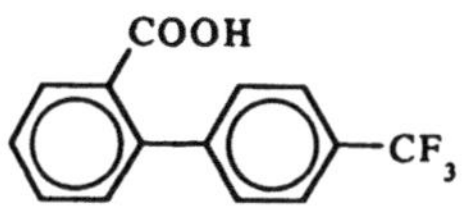 Xenalipin	Rat feed	13
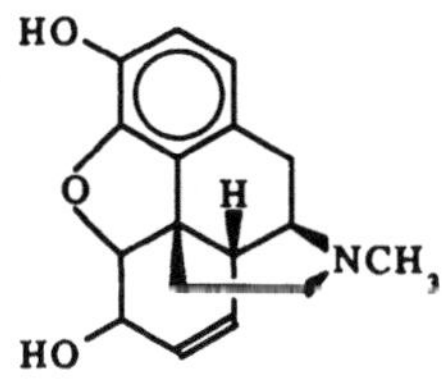 Morphine	Blood Placental tissue	19
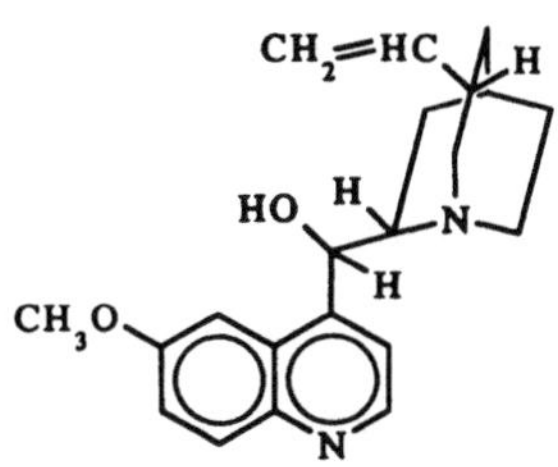 Quinine	Lemon grass Dogonyoro leaves	19
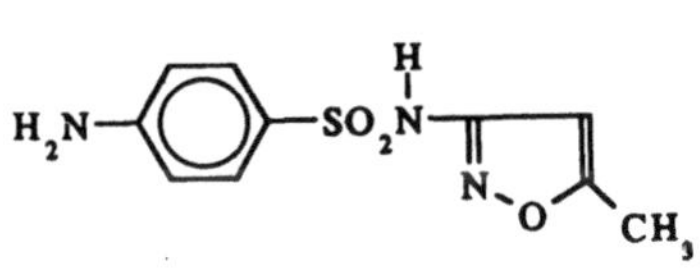 Sulphamethoxazole	Septra infusion intravenous solution	24
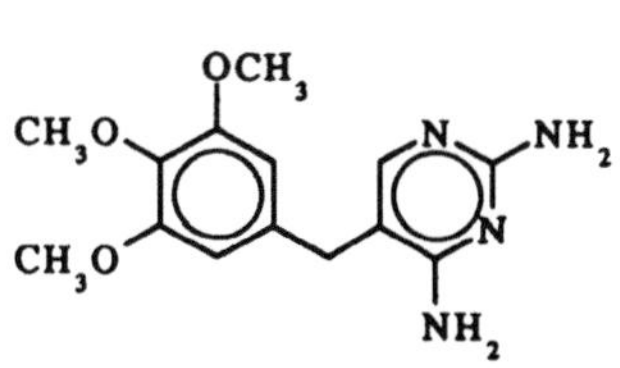 Trimethoprim	Septra infusion intravenous solution	24

Acknowledgement

The authors would like to acknowledge Dale Messer for his valuable insights into this project.

References

1. J. W. Dolan and L. R. Snyder, *Troubleshooting LC Systems*, Humana Press, Clifton, NJ (1989) Chap. 6.
2. S. R. Bakalyar, M. P. T. Bradley and R. Honganen, *J. Chromatogr.* **158** (1978) 277.
3. H. A. Norman and J. B. St. John, *J. Lipid Res.* **27** (1986) 1104.
4. L. A. Smith, H. A. Norman, S. H. Cho and G. A. Thompson, Jr., *J. Chromatogr.*, **346** (1985) 291.
5. H. C.- K. Chang and L. T. Taylor, *Anal. Chem.* **63** (1991) 486.
6. A. L. Howard and L. T. Taylor, *HRC & CC*, accepted for publication.
7. H. C.- K. Chang and L. T. Taylor, *J. Chromatogr.* **517** (1990) 491.
8. K. K. Unger and P. Roumeliotis, *J. Chromatogr.* **282** (1983) 519.
9. W. G. Engelhart and A. G. Gargus, *Am. Lab.* **20** (1988) 30.
10. J. B. Nair and J. W. Huber III, *LC/GC* **6** (1988) 1071.
11. H. Daimon, Y. Katsura, T. Kondo and Y. Hirata, *Proc. 12th Int. Symp. on Capillary Chromatography*, eds. K. Jinno, P. Sandra and T. Hanai, Convention Center, Kobe, Japan (1990) p. 456.
12. F. Andreolini and A. Trisciani, *J. Chromatogr. Sci.* **28** (1990) 54.
13. D. C. Messer and L. T. Taylor, HRC & CC, accepted for publication.
14. Y. Hirata and Y. Okimato, *J. Microcolumn Sep.* **1** (1989) 46.
15. M. A. Schneiderman, A. K. Sharma and D. C. Locke, *J. Chromatogr.* **409** (1987) 343.
16. H. H. Holton and F. L. Chapman, *Tappi* **60** (1977) 121.
17. V. Janda, G. Steenbeke and P. Sandra, *Proc. 10th Int. Symp. on Capillary Chromatography*, Vol. 1, eds. P. Sandra, G. Redant and F. David, Palazzo dei Congessi, Riva del Garda, Italy (1989) p. 457.
18. A. L. Howard and L. T. Taylor, unpublished results.
19. D. P. Ndiomu and C. F. Simpson, *Anal. Chim. Acta* **213** (1988) 237.
20. K. Sugiyama, M. Saito, T. Hondo and M. Senda, *J. Chromatogr.* **332** (1985) 107.
21. G. M. Schneider, E. Stahl and G. Wilkie, *Extraction with Supercritical Gases*, Verlag Chemie, Weinheim (1980) Chap. 1.
22. M. A. Schneiderman, A. K. Sharma, K. R. R. Mahanama and D. C. Locke, *J. Assoc. Off. Anal. Chem.* **71** (1988) 815.
23. M. A. Schneiderman, A. K. Sharma and D. C. Locke, *J. Chromatogr. Sci.* **26** (1988) 458.
24. L. J. Mulcahey and L. T. Taylor, unpublished results.
25. L. J. Mulcahey and L. T. Taylor, HRC & CC **13** (1990) 393.

Index

MIX
Papier aus verantwortungsvollen Quellen
Paper from responsible sources
FSC® C105338

FSC
www.fsc.org

If you have any concerns about our products,
you can contact us on
ProductSafety@springernature.com

In case Publisher is established outside the EU,
the EU authorized representative is:
Springer Nature Customer Service Center GmbH
Europaplatz 3, 69115 Heidelberg, Germany

Printed by Libri Plureos GmbH
in Hamburg, Germany